John H. Tice

**Elements of Meteorology**

Part II: Meteorological Cycles

John H. Tice

**Elements of Meteorology**
*Part II: Meteorological Cycles*

ISBN/EAN: 9783337276690

Printed in Europe, USA, Canada, Australia, Japan

Cover: Foto ©berggeist007 / pixelio.de

More available books at **www.hansebooks.com**

# Elements of Meteorology

## PART II

# METEOROLOGICAL CYCLES

BY

## JOHN H. TICE

*"Nunquam aliud natura, aliud sapientia dicit"*

SAINT LOUIS
Meteorological Research and Publication Company
1875

# PREFACE.

The opening sentence sufficiently indicates without further remark that this volume by itself is incomplete, and that it must be preceded by another, of which it is a mere episode, and in which general and more comprehensive principles are discussed. It may, however, be necessary to give an explanation of the anomaly of publishing the Second Part first.

In a sojourn in the Rocky Mountains, during the Summer of 1874, having leisure, I completed a work on Meteorology, based upon original researches, in which I have been engaged for over twelve years, the results of which I had been employed for the last five years in systematizing and elaborating, as time and opportunity offered. When completed I offered it to the Eastern publishers; but, with one exception, they all refused to look at it; and the exceptional case was a compromise, to get rid of importunity, by consenting to look at one Chapter. The following was the result: "We have examined the manuscript, and we think you have treated the subject with signal ability, and that the work is likely to prove interesting, and which will attract the general attention of scholars and scientific men; but for purely business reasons we have decided not to undertake its publication." These "business reasons," of course, were understood to be the same as those expressed by other publishers applied to: "Nobody reads Science, but everybody reads novels; if you have any thing of the latter we will give it favorable consideration." This we would take as a compliment to our system of public education if it stopped with the first proposition, "Nobody reads Science." But if the other proposition be just and true, that the mental cravings of the present generation, both young and old, are satisfied with the vapid and trashy romances and novels of the day, then it is the severest criticism of, and commentary upon, our system of public education, we have yet seen. If such be the result, then the less we have of it the better. After having devoted and spent the prime of my life in building up that system, I am loath to believe it; and take it to be that the public read novels because authors and the press do not furnish them any thing better.

But what shall we think of Science that nobody reads? "Ah! there's the rub." Science should be Wisdom; it should not only be Truth, but demonstrated Truth. Is this the case with what has been dignified as Science? We are afraid not. We have much to learn; and the Buddhists can teach us an important lesson; their god, Buddha, that is the Wise

One, becomes antiquated and worn out in 500 years, and then a new incarnation is necessary. Our wise ones never wear out or become antiquated; or, at least, we do not know when they have done so, for we never hesitate *in verba magistri jurare*. But is not that which we accept and swear by, effete, and belongs to the Dead Past? It certainly is without a vitalizing principle, for here we have the unimpeachable testimony of the experience of those who look at it in a business point of view, that Scientific works do not pay, for nobody reads them. Science, hence, is an antiquated god, and a new incarnation is necessary. Here is a glorious work for a Reformer; for he who snatches Science from the cloister, divests it of all speculation, and bases it upon indisputable facts, so that the commonalty can understand and comprehend it, confers a greater benefit upon Mankind that Prometheus did when he snatched fire from the gods and delivered it to mortals.

But I was going to tell why Part Second came to be published first. Not being able to get a publisher, and having no means of my own, a few enthusiastic friends volunteered to organize a Meteorological Research and Publication Company, to publish the work. But most of those who felt an interest had no means, and those who had means felt no interest. Consequently I found it necessary to do something to create a more general interest; and I resorted to the expedient of publishing one of my Meteorological discoveries, the one most likely to attract attention and excite a deep interest, and do it more expeditiously than any other, namely, Meteorological Cycles. In a newspaper article I stated the general principle, specified the dates of planetary equinoxes upon which it rests, and made forecasts of the weather, particularizing the kind of weather that might be expected, such as I invariably had observed, for eight years past, to attend these equinoxes. Three periods, the most strongly marked during the present year, were designated. One of these periods was a month off; the other five, and the last eight months. Two of these periods have passed, and with astonishing precision the forecasts, extraordinary as they were, have been more than verified. An interest was excited far beyond my expectations, and the public demand was for an immediate publication. But as the funds were coming in slowly, I thought it expedient to divide the original work by taking out so much as related to these Cycles, and publish it separately. Of course this necessitated a re-writing and a remodelling of the entire work, and has imposed an immense amount of labor upon me. To the friends who have assisted me, I am under lasting obligations, and return to them my heartfelt thanks.

JOHN H. TICE.

St. Louis, Mo., Aug. 6th, 1875.

# CHAPTER I.

*Meteorological Cycles. Their periodicity and cause. The length of the Great or Jovial Cycle determined. Its historical verification, Saturnian Cycle, etc.*

We have arrived at that point in the discussion where the review of the controlling Meteorological phenomena is complete. We have determined the nature and character of these phenomena, and we have ascertained their causes. We have given an intelligible explanation both of the phenomena and of the causes from which they spring, and of the laws by which they are controlled. Hence we know the conditions under which they occur, and the reason why they occur. But we do not yet know the periods at which the necessary conditions will supervene, and hence cannot tell the time when the phenomena will occur. Moreover, it is yet a mooted point amongst laggards in the science whether their occurrence is regular or irregular. This however is a vital point, for if Meteorological phenomena are irregular, then we will forever remain in the dark and at the mercy of chance; for as then we cannot foresee the future, so we cannot make provisions to meet its exigencies; because neither the time when, nor the nature of what the future may bring forth, can then be known to us. But if they are regular in their recurrence, then we know not only what to expect, but when to expect it, and provide for it in time, whether it be to take advantage of the opportunities it affords, or to guard against the damage it threatens.

Assuming then that we know all about the nature and constitution of rain and snow storms; all about cold and hot, wet and dry seasons; and all about winds, gales, tornadoes and hurricanes; yet this knowledge dwindles, in a practical view, into utter insignificance when compared with that which informs us of the time when these phenomena will be upon us, what their character and probable energy will be, and the length of their duration.

The regular recurrence of identical physical phenomena is now an admitted fact by all progressive meteorologists; but it is difficult to gather their notions as to the character of the cycle, whether it be of definite or indefinite length, and yet more difficult to ascertain what their ideas are as to its cause, or whether they have any notion whether it is caused at all. It is owing to the latter fact that they have made so little progress in placing Meteorology upon a firm and impregnable basis. It is evident that if it be once accepted as an undeniable truth, that phenomena repeat themselves in periods of a definite length, then the length of the cycle being accurately established, it will become evident that there must be a fixed cause producing this periodicity, and that cause can be discovered. It must be admitted that when such a cause is discovered, its nature and character can be studied; and that it can be ascertained under what conditions it manifests itself as cause, and at what time these conditions will recur. Now all that is necessary to enable us to determine Meteorological cycles is to acquire a sufficient knowledge on this point. When so much is achieved, we will be able to predict telluric and atmospheric perturbations with as much certainty as we now do eclipses of the Sun and Moon, or the occultations of Jupiter's satellites.

Meteorology is the science of physical phenomena, and until the true nature and causes of these phenomena are known, it cannot lay any claims to eminent domain even in its own proper department. Science is that which is known, and not that which is unknown. Therefore the first duty for its devotees is to prune it of whatever is unknown even if it leaves not a shred behind; for in what passes for Meteorology there is so much derived from the closet and so little from Nature, that nothing can be safely accepted upon trust. For self-protection against imposition we must apply the crucial test of facts to everything, and whatever will not bear the test, must unceremoniously be rejected. Meteorology is preeminently a science of facts, and of such facts as are easily observed and whose validity are as easily determined. It, hence, is inexcusable culpability to accept any thing upon trust when its truth can be so easily verified. There is no wrangling, but universal agreement on all questions that can arise in sciences resting upon indisputable facts; and there is also implicit acquiescence in all legitimate inferences drawn and logical deductions made from premises implied by the admitted facts.

A Science to deserve respect must be able to do more than account for the Past, or to explain the Present; it must be able to divine the Future. Hence it must comprehend fully that fundamental principle that underlies the department of the Universe embraced within its scope. It must not only know this principle, but it must know how it will act, and what will be the results under all the possible conditions that will successively be imposed upon it by the operations of the system of which it is a part. Hence Meteorology must know not only the cause of physical perturbations in the Earth and Atmosphere, but the exciting causes of them. Yea more, not only these, but the period and conditions when the exciting cause will awaken from a long repose into fearful activity. No one who has ever investigated the principles of Meteorology with a view of applying them to foretelling the Future, but has felt such knowledge indispensable. But when he has surveyed the field and looked at the labor necessary for its attainment, he has become dismayed at the long, tedious and difficult way to success. First the periodicity of the perturbing phenomenon has to be established. Secondly, the length of the cycle in which it repeated itself has to be determined; and thirdly, the fixed cause, its nature and mode of acting has to be discovered. The discovery of a Meteorological cycle,—the most clamant desideratum of the age,—seemed, hence to be postponed indefinitely, and only a remote possibility in the far distant Future. For half a century scientists have earnestly been laboring to discover such a cycle, but without success; yet every day the necessity for it becomes more evident and the demand for it more earnest and pressing. Professor Lockyer, an English astronomer, and renowned as a spectroscopist, expresses himself upon this subject as follows: "In Meteorology as in Astronomy, the thing is to hunt down a cycle; and if it is not to be found in the Temperate Zones, then go to the Frigid Zones, or to the Torrid Zone to look for it. If found then above all things, and in whatever manner, lay hold of it, study it, record it, and see what it means. If there is no cycle, then despair for a time if you will, but plant your Science on a physical basis."

Sir David Brewster, in his Meteorological report made to the British Association in 1840, says: "When these observations (those of Inverness and Hingussie) are compared with those made under my superintendence at Leith, with those at Plymouth from 1832 to 1840 at the expense of the Association under the able superintendence of Mr. Snow Harris, and with those at Padua, Philadelphia and Ceylon, we see very distinct traces of Meteorological laws of which no idea had been previously formed; and I have no hesitation in stating that when observations of this class are multiplied and extended they will lead to general results of as great importance in predetermining atmospheric changes as those which have enabled astronomers to predict the phenomena of the planetary system."

It is impossible to say what these "traces of Meteorological laws" were, for he neither states nor even intimates them. In this he committed a grievous fault. It was probably a mere light glimmering in the dark, a vision of the night that came to him in a moment of inspiration, but never assumed any definite form or shape. It may however have been that he discovered traces of covariation in observations made at such widely separated points. If so, these should have been followed up, for they would have led to the discovery of that great desideratum of the time, namely, Meteorological cycles. But this was unattainable without more definite observations than those furnished by the method then and still in vogue. To fix cyclical points, we must have individual phenomena, and not averages for a month, a season, or a year. Averages had and still have their value; they have led and still lead us to discover general laws and principles, but to discover special laws and special principles we must have special phenomena, that is, individual facts.

Observation on special phenomena, such as sunspots, solar physics generally, magnetic intensity in the Earth, electric tension both of the Earth and of the Atmosphere, auroras, earthquakes, Cyclones, rainfalls and terrestrial temperature, have been made for half a century, and some for nearly two centuries. Why these observations have not been more fruitful in valuable results is owing to the fact that each observer worked independently, and made the observations of his chosen phenomena a specialty, without ever dreaming that there was a corelation between all physical phenomena. After devoting a period to their work—equal to half the length of what the Psalmist assigns as the term of human life,—each observer came to the conclusion that his special phenomenon had a periodicity, and repeated itself in a cycle between ten and twelve years.

Sunspot observers such as Schwabe, Loewy, De la Rue, Wolf, Stewart, etc., differ as to the *maxima* of sunspots as much as half a year, and as to the length of their cycles deduced from their observations, two and a half years. Taking the average of their estimates, the length of the Great Cycle is found to be between eleven and twelve years. Cyclone observers in the East Indies, think they have discovered a cycle that completes itself in thirteen years; but Mr. C. Meldrum, of the Observatory at Port Louis, in the Isle of Mauritius, in the focus of the Cyclones of the Indian Ocean, shows conclusively from recorded facts, that the cycle has a period, the length of which is a fraction over eleven years. The late Professor Hansteen, of Norway, who made Magnetism a life specialty, fixed the cycle of Magnetic perturbation at 11.1 years. A recurrence of four or five cycles will show that this period is too short by over nine months.

Signally as each failed to establish the length of the cycle for his special phenomenon, yet the result of their joint labors, was an immense stride in progress; for their observations when compared showed that the maxima and minima of all the observed phenomena were coincident with each other, that is, they had the same periodicity, and in every respect were covariants. From these facts the legitimate inference is that concomitant phenomena, always appearing cotemporaneously, that are synchronous in their maxima and minima and of the same duration, are sequences of one and the same cause. Observation has verified this inference; for the period of maximum sunspots, is not more signalized for its auroras, magnetic and electric disturbances, and earthquake commotions, than for its copious rains, violent tornadoes, and destructive hailstorms.

The period of maximum intensity of the phenomenon was generally, because naturally, fixed upon as the term of the cycle. But all observers had also discovered that in this cycle there was another: a minor maximum but of far less energy than the greater. It was seen that the minor maximum divided the cycle into two nearly if not quite equal parts. Furthermore it was ascertained that between the major and the minor maximum in each half cycle there was a minimum corresponding in degree to the maxima. Other perturbations were observed to intervene, but which could not be referred to the Great Cycle because they happened at all points in it. They were therefore regarded as independent. Attempts were made to establish minor cycles by averaging the phenomena. But the averages obtained by lumping the phenomena, obliterated the individual facts, while they disclosed nothing. The result was hypothetical cycles of 28, 34, 56, 112, etc., days. But when tested by facts all of these, excepting that of 112 days, generally failed. At the proper place we will verify the 112 days phenomenon not as a cycle but as a section of a longer cycle. A longer cycle one of about 59 years was also discovered. This however is a conjunction of two cycles, two of one nearly equalling five of the other, as will appear at the proper place.

With so much devotion, and with such indefatigable zeal as the laborers in this field of Science have shown for the last half century, it is somewhat a mystery why their labors have not been crowned with better success in the attainment of the desired results. The failure must in degree, be attributed, either to their method of investigation, or to the hypothesis that directed it. So far as averaging phenomena was concerned, after averages had taught them all they were capable of teaching, namely in the most general way the existence of a cycle and where to look for it, their method was not only defective, but inadequate for the purpose of obtaining definite and specific results. Whether they had any hypothesis beyond the presumption of the existence of a cycle, it is difficult to tell. Now an hypothesis is a powerful instrument of investigation, but it must be a legitimate one, namely one that admits of verification. The hypothesis that suggests the direction of investigation, however, must always be subordinate to the investigation, and never must the investigation be subordinate to the hypothesis. No one can investigate without an hypothesis, nor, subordinate to one.

After having satisfied myself of the existence of meteorological cycles, about eight years ago I undertook to investigate their cause with a view of determining their length. As nothing can exist without a cause, synchronous and covariant phenomena regularly repeating themselves in cycles of uniform length, must have a permanent cause that is common to them all; and whatever that may be, and wheresoever located, it must be ascertainable and susceptible of proof. This was the only hypothesis I had, and the object of my investigation was to ascertain the length of the cycle and the cause of it, and to verify and demonstrate both. My first step was to collect all the material accessible to me, consisting of the records of observations made upon physical phenomena such as auroras, sunspots, Cyclones, rainfalls, earthquakes, etc. After carefully sifting their dates, by lumping them I obtained 11.83 years as the average length of the cycle indicated by the observations. As this period corresponded so closely—within 11 days of the Jovial year, I projected the hypothesis that Jupiter in some unknown way was the cause of the perturbation. I then made a historical record of all the marked periods of disturbance, such as earthquakes, auroras, Sunspots and Cyclones for 2500 years. With these facts at command I proceeded to test the hypothesis that Jupiter was the cause of the perturbation. It must here be stated that all the observations show that the period of perturbation extends over about three years, manifesting itself often two years in advance of the maximum and a year and a half afterwards. I did not however admit any facts as verifying the hypothesis excepting those that came within twelve or sixteen months. The reason for the length of the Jovial disturbance is, that Jupiter moves very slowly in his orbit only about 30 degrees in a year. The year 1859 it is generally admitted was the year of maximum disturbance. It was therefore taken as the standard\* and the time intervening between it and the date of the phenomenon, was divided by 11.86 years the length of the Jovial year. It is well known that scarcely a year passes without a few sporadic earthquakes, that there are always some sunspots, occasionally a faint aurora, and more or less violent Cyclones, but these are few, far between and feeble, when compared with the incessant and intense energy of the phenomena occurring in the perturbed

---

\*NOTE.—Since then the equally strongly marked cycle of 1871 has occurred. In the subsequent pages I have taken the latter for the standard of comparison.

cycle. But of nearly two hundred historical phenomena there were but three intensely and strongly marked that did not coincide within limits with the Jovial cycle. Many of them corresponded to the very day. The three exceptional ones, I subsequently ascertained belonged to a Saturnian Cycle. At the proper place a sufficient number of facts and their correspondence will be given to verify the hypothesis that Jupiter is the cause of the perturbation. I have hence named the period the Jovial Cycle. Having satisfactorily demonstrated and verified the proposition that Jupiter is the cause of the atmospheric, telluric and solar perturbations that occur once and in a modified form twice in every one of his orbital revolutions, it remained for me to ascertain the cause of this disturbance.

In the preceding part of this work I have demonstrated that winds, rain, snow and hailstorms, Cyclones, auroras, earthquakes, in fine all telluric and atmospheric phenomena are electric; and that under what may be considered the normal condition of the Earth and the Atmosphere the Electricity necessary to their production is constantly being generated but with varying energy. Under what may be considered an abnormal condition imposed upon them by the Jovial Cycle, the character of these phenomena is not changed, but their energy is only terribly intensified. Hence it follows that in whatever way Jupiter may affect the Sun and through the Sun the solar system, the result upon the Earth and the Atmosphere is an enormous increase of electric intensity. What is it that is taking place in Jupiter, that he produces this effect at two particular and opposite points on his orbit? This was a question I propounded to myself for solution, and I immediately set about solving it, but fully two years elapsed before I attained to a satisfactory result.

Since the maxima of telluric and atmospheric disturbances are synchronous and covariant with solar disturbances as manifested by sunspots and immense solar explosions, therefore the proximate cause of telluric and atmospheric disturbances must be the Sun. But as solar perturbation invariably is synchronous with Jupiter's passage through a particular point on his orbit, so Jupiter's position on the particular point, must be the cause of these periodical disturbances in the Sun, and consequently on the Earth, and most probably on the entire solar system. Sunspot observers had pretty generally agreed that the maximum of sunspots in 1859 occurred about 1859.90. Calculation placed Jupiter at that period at 101° of the Ecliptic. But as no cause could be discovered why he should at this point exert such influence, investigation in this direction for the time being was suspended. True, that point is in the plane of the Milky Way and also within about three degrees of his ascending node; but evidently these were accidental, and had no significance, since no possible reason can be assigned why Jupiter's ascension above the plane of the Earth's orbit, or in the plane of the Milky Way, could have any effect on the Sun. In this way Inquiry was pushed in every direction but returned empty-handed. Speculating upon what possibly might be taking place in Jupiter was even less fruitful of results. The conclusion finally arrived at was this: We know too little of what is taking place in Jupiter to serve us in unraveling his mysterious influence on the Sun. If we were inhabitants of that majestic orb, we would know all the facts, and hence could assign the reasons why he periodically sends out an influence which thrills through the solar system, and is felt to its remotest points. But as we are not, our only resource is to fall back to our own planet and learn whether analagous facts are occurring there; for what is true in Theology is true in Science.

"Through worlds unnumbered though God may be known,
'Tis ours to trace Him only in our own."

Jupiter's influence upon the solar system is electric. Hence if his orbital position at the time influences his own electric condition, then the orbital position of the Earth and of any other planet must have a like effect upon them under similar circumstances and conditions. Whatever may be the case with other planets, it is undeniable that the electric condition of the Earth is affected by its relative position to the Sun. If annual telluric phenomena are separated into two classes, namely, equinoctial and solsticial, then the equinoctial will be to the solsticial as three is to one. For example, take the tropical hurricanes, and nine out of ten occur in the Northern Hemisphere in the months of August, September and October, at or near the Autumnal Equinox; and the same proportion occur in the Southern Hemisphere in February, March and April; at or near the Vernal Equinox.

The same is true of auroras and earthquakes. It is indisputable that the Earth and the Atmosphere undergo an electric pertubation at our equinoxes. All physical phenomena indicate this; the tides then run higher, the barometer ranges lower and electric tension on the Earth and the Atmosphere is then greater than at any other season of the year. It is established by observation in both the Northern Hemisphere and the Southern, that magnetic perturbations have their maxima at the equinoxes, and their minima at the solstices.

Mr. Brown, in his report on the magnetic observations at Makerstoun for 1846, as quoted by Gen. Sabine, states the results, with respect to the frequency and magnitude of magnetic perturbations in the different months of the year to be as follows: "The mean value of perturbations is a *maximum in the equinoctial months*, a minimum in the summer months, and a proximate minimum in the winter months." Gen'l Sabine states that at Hobarton, Tasmania, "the proportions of the frequency and of the magnitude of these perturbations in each month relatively to those observed in a year, come out, a minimum in the winter months, *a maximum in the equinoctial months*, and intermediate in the summer months." These results are identical: for the Winter months of the Northern Hemisphere correspond to the Summer months of the Southern; and *vice versa*. It is also established by observation that sunspots have a marked maximum at the same period.

The equinoxes are especially marked by atmospheric perturbations, frequent and extreme oscillations of the barometer, violent gales, furious tornadoes and a "swing" of the periodical winds, such as the Monsoons. At one equinox Summer appears and leaves at the other. Winter appears when Summer leaves, and leaves when Summer appears. Each brings and takes its peculiar phenomena with it. Summer brings with it a low barometer to replace the high barometer that has prevailed all Winter over continents in that hemisphere. Winter returns and brings with it a high barometer to take the place of the low barometer that swayed the continents during Summer. Hence a dynamic electric pole takes possession of continents for the Summer; and the static that ruled them during Winter migrates to the other hemisphere to bear sway over continents there until the Sun returns again to that side of the Equator. In consequence of these changes all the winds on the surface of our planet, are more or less changed in direction, and many of them reversed.

The causes of these great and extreme changes are many; and each bears its proportionate share in the influence that produces the varied and general perturbation prevailing at the time. The facts that a hemisphere in which solar influence has been feeble for six months and over a large fraction of which that influence has been null during the same time,—being wrapped in uninterrupted darkness,—is now falling directly under the full power of the Sun, and that the other hemisphere which for the same length of time has felt the full energy and effect of solar influence is now being withdrawn from it, contribute largely and powerfully in exciting the general disturbance; for all elements of physical disturbance, thermal, electric and magnetic, accompanying the Sun across the Equator. But there are remote, cosmic causes that contribute to the same end. In consequence of the inclination of the Sun's axis to the plane of the Ecliptic, the South magnetic pole of the Sun at one equinox, and his North magnetic pole at the other, are pointed more directly toward the Earth than at any other points on its orbit. Consequently the Earth at its equinoxes feels the full force of the Sun's magnetic influence. Since it is probable that the magnetic poles of the Sun, like those of the Earth, do not coincide with the poles of his axial rotation; hence if this is the case, each solar pole is equivalent to a moving magnet to all the planets, and therefore a powerful generator of Electricity in them; and especially will the Earth at the equinoxes feel their most powerful effect. The inclination of the Sun's axis to the plane of the Ecliptic is according to Herschell, $7°, 20'$. Other authorities say a few minutes less. The inclination is nearly in the plane of the equinoctial colure, or Equator. Consequently at the equinoxes—or more accurately ten days before—the plane of the Sun's equator makes its greatest angle with that of the Earth. Hence the nodes of the Sun's Equator, that is the plane of his axial rotation, must cut the plane of the Earth's orbit at some point, or rather two points. It does so at the points occupied by the Earth on the 11th of June and 12th of December. This is evident from the paths of sunspots; on or near the 11th of June or the 12th of December, their paths are straight lines, showing that the Earth is then in the plane of the Sun's Equator. On or near the 11th of March or September their paths are sections of elongated ellipses. At the vernal equinox, a sunspot ascends above the Equator, at the autumnal it sinks below it. All rotating bodies, from a disc to a sphere, generate electric currents at right angles to the axis of rotation. On a disc the Electricity so generated accumulates—from the effects of self-repulsion—on the edge of its circumference. On a sphere from the same cause it collects on the equatorial belt. Electric charges are always driven by repulsion to the remotest points from the centre. Hence electric currents carry Matter away from the centre towards the circumference; and electric Repulsion, and not centrifugal force—as the mechanical theorists suppose,—has given the form of oblate spheroids to all the planets.

The mutual disturbance of the Sun and a planet at the time of the latter's equinox, is explicable upon well established electric laws. Viewed upon electric principles, a rotating disc and a rotating sphere, as far as the generation of Electricity is concerned, are identical; and how a pair of each will mutually affect each other is exemplified by two circulating currents with fixed centres but flowing in different planes which are free to assume any position. If they circulate in the same plane, they remain at rest. If they circulate in different planes, they mutually exert such an influence upon each other as to change their planes until they coincide. Consequently it is evident that when they circulate in one and the same plane, they are in equilibrium, and exert no disturbing effect upon each other. Hence it follows that when they are out of the same plane, that is, when their planes of circulation intersect each other at the greatest angle, possible, they exert the greatest disturbing influence upon one another. It has been ascertained that the tendency of this influence is mutually to swing each other around into the same plane. Now from the inclination, and from the direction of the inclination of the Sun's axis of rotation to the plane of the Ecliptic, and from the inclination of the Earth's axis of rotation to the plane of its orbit, it follows that the plane (equator) of the Sun's rotation, and that of the Earth, make their greatest angle with each other at or about the equinoxes. And since, as we have seen, the equatorial belts of all spheres, and especially of spheroids by the operation of a universal law, become more intensely charged with Electricity than any other points on their surfaces, hence whatever electric influence they are capable of exerting upon each other, must be at its maximum when their planes of rotation make the greatest angle with each other.

Astronomy establishes two points, namely, that at the equinoxes the Earth's Equator has its greatest obliquity to that of the Sun; and that at or near the solstices it has its least, since the plane of solar rotation then passes through the Earth. Electric laws exact that under such conditions electric excitement should be at its maximum at the equinoxes, and at its minimum at the solstices, and observation has confirmed that the facts correspond with the obvious deductions of Reason.

An investigation of the facts of our own Globe has disclosed phenomena that Reason suggests must be analogous to those that must be taking place in Jupiter, when he sends forth periodically that mysterious influence that affects the entire solar system. These terrestrial phenomena we have ascertained occur invariably when the Earth passes the equinoctial points. We have seen that physical changes, opposite in character, are taking place in both its polar hemispheres which affect their thermal, their electric, magnetic, pneumatic and hygrometric condition. Behind these obvious changes we have discovered cosmic causes and laws, whose inevitable influence under the existing physical circumstances is to intensify these phenomena. Analogy justifies and legitimates the application of these physical facts and the principles in which they have their being, not only to Jupiter, but to all the planets of the solar system, for the same laws and causes existing there, must under similar conditions effect the same results.

It is an old fact, well known but not comprehended, that in a system of insulated bodies we can neither increase nor diminish the electric tension on any one of them, without changing in the same degree the electric tension in all. Applying this principle to the solar system, if from internal or extraneous influences an electric charge is evolved on a planet, or is either augmented or diminished, then that planet not only disturbs his neighboring planet, but all his co-planets and even the great Sun himself.

This fact discloses the inspiring truth that the Universe is far more symmetrical in its structure and far more delicate in the adjustment of its parts, than our crude and inelastic theories permit us to imagine. Myriads of facts exist that teach us this ennobling and sublime truth; but we are too blind to perceive the phenomena, too deaf to hear the voice of Nature addressed to us through them, and too ignorant to understand their meaning. The difficulty in attaining to a true conception of the solar system and of the structure of the Universe, is not because that system is abstruse, for it is perfectly plain and comprehensible, but because we are afflicted with a mental nightmare, the Mechanical Theory, that presses upon us and renders us mentally helpless. Its overshadowing notion is that the Universe is a rigid machine, operated by mechanical powers; hence it not only checks but represses thought in the direction that would lead to the conception of a pure dynamical theory of the Universe with a structure and adjustment of its parts such as to be normally a perpetual motion under the influence of a system of Cosmical Forces.

There are two facts that are undeniable; which are not only necessary to establish our theory but they render it incontestable. The first fact is that the Earth and the Atmosphere at the equinoxes always undergo an intense electric disturbance; and the second is that the telluric disturbance extends to and affects the Sun; and hence even though she may not directly communicate her disturbing influence to the other planets, yet she does so indirectly through the Sun. Since this action of the Earth is so evident as to be incontestable therefore by an extension of the principle it must follow that Jupiter, with a volume 1491 times greater than the Earth, when undergoing an electric perturbation must obviously affect the whole solar system.

It is not presumable that the physical relations of planets to causes of disturbances are identical in every respect with those of the Earth. In fact we know they are not. The relations of one planet in regard to the Sun may be such that some of the elements of disturbance may have less energy than in other planets, and other elements have more; but there can be no doubt that all the elements of perturbation are physically interwoven with, and inseparable from the planetary system. Physically we know that they do disturb each other, for it takes place under our eyes. At one time we see they retard, and at other times accelerate each others velocities, and are constantly forcing each other to make deviations from their regular orbital path. Since it is indisputable that they influence each other in their paths through Space, why should it appear incredible that they affect each other dynamically, either directly or through the mediation of the Sun? We predict that when Science has so far advanced, that it can and will observe the phenomena resulting from what are now supposed to be mechanical influences, it will be ascertained that they are not mechanical but dynamic, and are produced by electric induction, repulsion and attraction; and that more or less an electric disturbance then takes place in the planet. Since the orbit of each planet inclines at different angles to the plane of the Ecliptic (the Earth's orbit projected into Space) therefore no two planetary orbits can make the same angles with the solar equator, and hence all must experience an electric perturbation when they pass the points on their orbits where the plane of their axial rotation makes the greatest angle with that of solar. Then again, no two planets have the same inclination of their axis of rotation to the planes of their orbits. Hence while some of the elements of perturbation may affect them less than on the Earth, others may affect them more, but evidently none are null.

We will note a few modifying influences operating in Jupiter, some of which must moderate while others must aggravate his perturbation at the critical periods. The fact that his polar axis makes only an angle of $3°$ with the perpendicular of the plane of his orbit, must modify the violence of the transitions of his seasons, as compared with those of the Earth. In fact Summer and Winter themselves cannot be as strongly marked in Jupiter as on the Earth, from the comparatively small obliquity of his axis of rotation as compared with that of the Earth, notwithstanding his greatly flattened poles, his polar diameter, according to Müller, being one-fourteenth less than his equatorial, while that of the Earth is only one two hundred and ninety-ninth less. On the contrary his great polar compression, his enormous bulk, and the great velocity of his axial rotation immeasurably intensify evolution of Electricity, and consequently its tension on his equatorial belt where it naturally collects.

The physical aspects of Jupiter, if closely observed, must afford evidence of the intensity of the electric charge upon him and upon his atmosphere during his equinoctial perturbations. Reason suggests this; but here again, as at so many other points, when we logically push inquiry to and beyond the outposts of Knowledge, we find no systematic observations. But if the facts be—as this legitimate inference from the theory suggests,—then they must be too obvious to have escaped observation entirely for over two hundred and fifty years, which is not possible, because ever since Galileo, in 1610, discovered his satelites, Jupiter and the Jovial system have been under constant observation and study.

On the physical aspects of Jupiter we find only observations recorded on periodical spots, on his belts, and on variations in the size and color of the belts. The dark spots evidently belong to the body of the planet since they never change in position and pass across the disc parallel to his equator and with uniform velocity. In fact the time of the axial rotation of Jupiter has been determined by these spots. Cassini, in the winter of 1665 and 1666, observed a spot which afterwards reappeared seven times between that year and 1708. It generally remained visible about a year, and then disappeared for five years. Or in other words, the interval between Jupiter's equinoxes is six years, lacking only twenty-two days. The frequent reappearance of this identical spot leaves no doubt that it is permanent on the body of the planet; and that its disappearance is owing to the interposition of clouds in the atmosphere of Jupiter. I find that Brewster, more than forty years ago, arrived at the same conclusion. In 1785 and 1786, while Jupiter was at his perihelion, Schroeter observed several spots which were black and round. These spots however must not be confounded with those that result from the breaking up of the belts, which distribute themselves over the whole face of the planet, and which revolve with varying rapidity. These latter spots have different shades of light, from dark to very bright.

Sir John Herschel supposed that even the belts were the dark body of the planet, and assigned as a reason that these belts never come up in all their strength to the edge of the disc, but gradually fade away before they reach it. I cannot concur in this opinion, because the belts never are persistent either in *in esse* form, size, position or color. Sometimes they are absent. Generally they consist of straight lines forming continuous belts around the body of the planets. Sometimes however the belts are discontinuous, and at other times the lines are curved and irregular, but lying in belts all of which are parallel to the planet's equator. The belts sometimes break up, forming transient spots, which may spread over the whole face of the planet, and which move with varying velocity across its face. The facts seem conclusive that they cannot belong to the surface of the planet, but to its atmosphere.

Different opinions have been held as to the cause of these spots and belts. Some supposed they were clouds and openings in the planet's atmosphere. So far as the belts are concerned we have no doubt that this is true; but it evidently is erroneous as to the persistent spots which are always the same in form, position and size, appearing at regular intervals, and after having been visible for nearly a year, disappear, and are not seen again for five years. Others regard the spots and belts as indications of great physical revolutions perpetually agitating and changing the face of the planet. Had they said physical perturbations that affect the planet's atmosphere we would think them about right. But when they ascribe these appearances to incessant physical revolutions that are continously changing the aspects of the planet, they propound an hypothesis so palpably absurd as to require no refutation. These appearances can be explained by far more rational causes than perpetual vexings of the planet by hurricanes, waterspouts, inundations and earthquakes; in fact, they can be explained upon natural causes so obvious as to be self-evident.

We have already adverted to the fact that the spot first observed by Cassini, reappeared about every six years, remained visible for from five to ten months, and then disappeared, but invariably returned in about five years. This is very significant, because it coincides precisely with the interval between the equinoxes of Jupiter, namely, 5.93 years. Schroeter, in the Winter of 1785 and 1786, observed the same spot discovered by Cassini 120 years before. Schroeter incidentally remarks that the planet was then at his *perihelion*. In 1834 Mädler and Beer saw and figured the same spot and several others, all of which remained visible from the 4th of November, 1834, till the 18th of April, 1835. Now it is remarkable that at the time these spots were seen Jupiter had just passed his perihelion. When we examine the position of Jupiter at the time Cassini discovered the spot, and his positions at the periods of the subsequent reappearances for 43 years, the period of its reappearance is invariably found to be when the planet was either at or near his perihelion, or aphelion. The perihelion and aphelion points, let it be remembered, are like the same points on the Earth's orbit, intermediate between the equinoctial points, and hence mark the planet's greatest tranquility. Our theory is that at his equinoctial points, Jupiter suffers physical perturbations both in its body and in its atmosphere, probably more intense than our telluric disturbances at our equinoxes. These will cause similar atmospheric and physical paroxysms in Jupiter as our equinoctial disturbances do, namely, electric and magnetic storms and earthquakes in the body of the planet; and in the Atmosphere violent tornadoes and hurricanes, accompanied with terrible electric explosions and heavy rain and hail storms.

Since the equinoctial perturbation of Jupiter lasts about three years, namely, one and a half year before and nearly as long after he passes the point, hence during this time his atmosphere must be more than usually surcharged with clouds that will form continuous belts around his body, impenetrable to the sight; and hence nothing on his surface during this period can be visible to us. But at his perihelion, and also at his aphelion—each midway between the equinoctial points—the perturbation being at its minimum, cloud formation must also be, and hence his atmosphere serener than at any other time. It therefore follows, that if any thing on his surface is visible to us at any time, it must be at the period when he is either at or near his perihelion or aphelion. Now it so happens, that always when Jupiter is at or near these points that these persistent black spots are seen on his disc; and seen only so long as the theory assigns for the duration of the interval between the disappearance and return of his cloudy season. If this be not considered as sufficient proof, and a confirmation and verification of the truth of the theory, then it must still be admitted that it is at least a most remarkable coincidence.

Campani saw two luminous belts across the disc of Jupiter; but I am unable to find the date of the observation. In the Winter of 1787 and 1788, and within eight months of the passage of Jupiter through the equinoctial point, producing the greatest disturbance, Schroeter observed that the equatorial zone belt had assumed a gray color bordering upon yellow. The equatorial belt was flanked on both sides by dark belts. Each of these on the polar side were flanked by white luminous belts resembling those described by Campani. These belts underwent a number of changes during an observation extending over several months. The dark belts sometimes suddenly increased in size. The luminous belts also suffered many changes, increasing sometimes suddenly until they were one-half times larger than their original size, then growing narrower. In 1871, when the last Jovial equinox occurred, European observers saw not only the usually dark gray belts change to yellow, then to orange, but for some time even to fiery red. We hence find that our inference, that the physical aspects of Jupiter must show the effects of the perturbation with which our theory supposes he is then affected, is verified; although no direct observations had been made to ascertain and establish that fact; yet we consider the verification more satisfactory and complete because the observations were not special, but if anything accidental. Moreover, the facts our investigation has elicited, do more than verify the inference that the effects of the perturbation must be visible on Jupiter himself, for they incontestably establish the existence of the perturbations, and prove that they occur at the equinoctial points of the planet.

We have already stated that we assumed the perihelion and aphelion points of Jupiter's orbit to be intermediate between his equinoctial points. Astronomy is silent as to planetary equinoxes; those of the Earth excepted. When I was satisfied that the equinoxes of Jupiter were the causes of the two marked periods of disturbances in his year, I felt disappointed and discouraged when I found the books silent on this now to me vital point, because it was the turning point either to disprove or confirm the projected theory. Should it confirm the theory, then it would settle a momentous question in Meteorology and Science; and one that is of inestimable importance to the destiny of the Human Race. So I thought and felt, because of my firm convictions of its truth and unshaken confidence that it would be confirmed by the crucial test of facts. I however did not despair. Analogy had suggested to me the Jovial equinoxes as the causes of perturbation. A thorough investigation had established that the facts exactly corresponded with the hypothesis; and when the points on the orbit that Jupiter occupied at the period of disturbance were determined, it was found they were opposite points. Hence it only remained to establish that these were his equinoctial points.

The first thing was to establish the solsticial points, upon which the books are equally as silent as upon the equinoctial. But as telluric analogy had served my purposes well so far, I relied upon it to furnish me both the solsticial and equinoctial points. On the Earth's orbit I found the perihelion and aphelion points closely coincided with the solsticial points; and the latter exactly coinciding with the points of the Ecliptic where the Sun apparently, but really the Earth, is highest above or deepest below the Equinoctial Line. I hence inferred that this might be the case with Jupiter; and if so, that the law might be general. But how was this to be ascertained?

Unquestionably Jupiter's solstices are at the Jovial Tropics, that is at the points where the Sun having made his greatest ascension North or declension South as seen from Jupiter, returns towards the Jovial equinoctial. But no works—at least none accessible to me—furnish any direct information on this point. The inclination of Jupiter's orbit to the plane of the Ecliptic is given, but not the point towards which it inclines. There is nothing hence to indicate the points of the tropics of Jupiter. But by following the suggestion that they would be found where the plane of his orbit is highest above, or deepest below the plane of the Ecliptic, it was inferred that the Jovial tropics, and consequently solstices, might be approximately ascertained. Inquiry was instituted in that direction: and it was ascertained that the Tropics or solstices of Jupiter were at the points inferred, and that they exactly coincided with his perihelion and aphelion; and hence of course his equinoxes must be 90° from either of these points.

When the hypothesis first suggested itself that Jupiter was the cause of the physical disturbances, such as sunspots, auroras, hurricanes, earthquakes, etc., which, as had been established, had two maxima—a major and a minor—in the Jovial year, my first impression was, that it was owing to Jupiter's position in Space; and hence I deemed it of the highest importance to fix these positions. Taking the record of all the phenomena I soon determined approximately the date of their maxima at each period. I next calculated the positions of Jupiter to correspond with these periods. I determined that one point was on or near 101° of Celestial Longitude and the other on 281°. As these were opposite points I felt much elated and became confident of success. Examining these positions relative to Space to ascertain what might be the mysterious influence exerted upon Jupiter at these points of his orbit so as to cause him to perturb the whole solar system, I found he was at both of these points in the plane of the Milky Way. I hence for some time had an hypothesis, that the influence that affected him came from Space, and in some incomprehensible manner was exerted by the Milky Way.

But as this hypothesis was incapable of verification I finally discarded it; for since it could not be verified, it could neither increase Knowledge nor advance Science. It could lead to interminable and indefinite speculations, but to nothing clear, definite and tangible. Hence after I had adopted the equinoctial theory as more reasonable and probable, I always felt that if the points of the Jovial equinoxes could be determined, and if they were found to coincide with these points in the plane of the Milky Way, it would be considered as a demonstration of the theory, and show that the locality in the plane of the Milky Way was merely accidental.

Now Jupiter's perihelion and corresponding solstice, is in Celestial Longitude 11° 45', 33"; consequently, his first equinox—the one that produces the maximum disturbance—must be 90° from the solstitial point; that is L. 11°, 45', 33" *plus* 90°, or L. 101°, 45', 33"; and his second equinox at L. 101°, 45', 33", *plus* 180°, or 281°, 45', 33". It was with inexpressible delight and gratification, mingled with astonishment, that I beheld these results coming out with mathematical precision.

Two astronomical points on Jupiter's orbit had been established by a careful analysis and collation of facts, not to subserve any theory, but to test the truth of every theory that might be proposed. Months, and even years had elapsed, when a theory was suggested based hypothetically upon the occurrence of physical events when the planet passes two definite points on its orbit in Space. Records of facts and phenomena were collected and compared with the periods; they were found to correspond in time and place with the demands of the theory. But this was not deemed conclusive, because there was no evidence that the astronomical condition existed at that point. No facts were known that established the assumed astronomical condition there or any where; but by a course of deductive reasoning, another astronomical point is determined, located one-fourth the immense distance around Jupiter's orbit from the point whose astronomical relation is to be determined. By the aid of this fact the exact location of the point in question could be calculated, and behold calculation fixes it at the exact points the theory had assigned to it, and where facts had located and proven it to be. So astonishing was the result that I could scarcely believe it. The labor of years had now been crowned with success, and it seemed to me as though the result had been attained more by divination, than by the cool and dispassionate deductions of reason.

Before proceeding any farther, it is necessary to remark that I do not claim astronomical exactness for the planetary equinoxes as I shall give them, nor do I expect that they will in all cases be found to be precisely as given. All that I claim is that, they are approximations arrived at from general principles, the only data at my command. My purpose is to prove that planetary equinoxes affect, and I might say, determine the meteorological phenomena of our Globe. I could not succeed in my purpose unless I knew, at least approximately, the points on the orbits of the several planets where their equinoxes occur. If with these approximations I succeed in establishing this great truth, Astronomy will see the importance and necessity of determining the exact points where the equinoxes occur, that their periods of recurrence can be calculated. In Jupiter the perihelion and aphelion points on his orbit, and his solstitial points exactly coincide, judging by the plane of the Ecliptic. Hence probably Jupiter's equinoxes as given will be found nearly if not quite exact. But the perihelia and aphelia points on the orbits of all planets may not always, like those of the Earth, exactly coincide with the solstitial points. In fact, judging by the paths of their orbits as compared with that of the Ecliptic, the equinoctial and solstitial points in some vary by a few degrees. Hence their equinoxes may vary a few days in their occurrence from the dates given by me. In such slowly moving planets as Mars, Jupiter, and Saturn, a few days and even weeks will make but little difference, but with Venus and Mercury, especially the latter, the period should be known to a day. Of Vulcan, the other interior planet, too little is known to claim exactness In fact, so far as my information goes, it has only been seen twice. Once by its discoverer, M. Lescarbault, March 26th, 1859; and by myself, Sept. 25th or 26th, 1859, in its transit across the Sun, in the forenoon. Without reflection, I supposed it to be Mercury; but several weeks later, when it occurred to me that a transit of Mercury in September was an impossibility, I could not recall and fix the exact date.

In 1871, Jupiter passed his ascending node on the 21st of August. This point is on L. $98°$, $48'$, $37''$, and his nearest equinoctial point, as we have seen, is L. $101°$, $45'$, $33''$. These two points differ only $2°$, $56'$, $56''$. Since his daily motion is $4'$, $59''.3$, therefore the difference between the time of his ascending node and his equinox is thirty-five days. Hence the equinox of Jupiter occurred in the year 1871, on the 26th of September.

In 1869 I determined to watch closely the phenomena of the impending perturbation expected to attain its maximum in 1871. I hence opened a record on which I entered the date of all phenomena reported to have occurred throughout the Earth, from the 1st of November, 1869, to 1st of November, 1872. The object was general; but special points were kept in view. Amongst these latter may be mentioned: (1) To ascertain the kind and character of the prevailing phenomena during the period. (2) To fix the maxima; and, (3) To establish their corelation.

Since the theory exacts that the maximum disturbance must occur at or near the Jovial equinox, I will quote the record for fifty days after the happening of the critical period, namely, September 25th, 1871. This is by far the period of greatest energy recorded during that year, but the record shows that during the whole year the phenomena—and especially those at the Venusian periods of which this is one—were incomparably more energetic than at any anterior or posterior period in the three years covered by the record. During 1871 two other periods of intense energy occurred; namely, one in February and March, owing to the disturbances of the Venusian* equinox of March 5th, and that of the Terrestrial of March 21st, being superimposed upon the Jovial which had already acquired great energy as was manifested by the frequency of earthquakes, the increase of sunspots, the brilliancy of the auroras, and electric and magnetic disturbances in the Earth. The other intense period of disturbance in June and July, was occasioned by the recurrence of a Venusian equinox on the 25th of June. An abstract of the record showing the phenomena at these critical periods will be given at the proper place.

I will however remark that the extreme perturbation exhibited by the phenomena between the 5th of August and the 25th of October, was not owing purely to Jovial influence, for within this period fell no less than nine planetary equinoxes, namely, Mars on August 5th, Mercury on September 8th, Earth on September 21st, Jupiter on September 26th, Venus on October 15th, and four Vulcanian namely, Aug. 13th; Sept. 5th and 28th, and October 21st.

I will here call attention to one of the most important facts of these extraordinary cyclical perturbations, which would strike us with amazement, if we did not see it to be natural. The energy of the equinox of any planet is intensified when that of another occurs at or about the same time. The reason of this is obvious, because the energy then manifested is the aggregate energy of both. I have verified this by calculation—where I had the exact dates—of phenomena considered so remarkable as

---

*Note.—No apology is required for coining this new word, 1st, because it is more intelligible and far more elegant than the old adjective derived from the Latin declension of the noun Venus; and, 2nd, because the old adjective has been so exclusively appropriated as to be inseparable from a foul and loathsome disease, so as to render its application to anything else improper and impolite.

to pass into history. Some of these occurred six, seven and eight centuries ago, yet I invariably found them to happen at periods of extraordinary conjunction of planetary equinoxes. Although the excitement of a Jovial equinox endures nearly three years; and manifests its presence constantly in some inobtrusive form, such as electric and magnetic disturbances in the Earth and in the Atmosphere, rapid and extreme oscillations in the barometer, heavy rainfalls, frequent auroras, and general seismic disturbances, yet sporadic paroxysms occur at a time when these phenomena display more than usual energy and violence. These paroxysms are noticed to be intermittent, and when investigated, are ascertained to be brought about by the superimposition of the ordinary excitement of some other planetary equinox upon the general disturbance then prevailing. For instance, a Vulcanian equinox ordinarily passes without causing phenomena so remarkable as to attract attention, yet when it occurs from eight to fifteen days after a Venusian equinox, the phenomena are often terrible. As examples, the Iowa Tornado of May 22d, 1873, and that which destroyed Tuscumbia, Alabama, November 22d, 1874, afford striking illustrations. Keeping these facts in mind, we can readily understand the violence of the phenomena in the record which we will now quote.

### RECORD.

Auroras are recorded in Month of August, 1871, as follows: 5th, 9th, 10th, 11th, 12th, 13th, 15th, 16th, 17th, 19th, 21st, 23d, 24th, and 31st.

REMARKS.—That on the 16th was very bright, and the prevailing color orange red. It was synchronous with the great sunspot; and that of the 21st, which was deep red and extremely brilliant, was synchronous with the terrible cyclone and earthquake at the Isle of St. Thomas.

Continuous sunspots during the month. From the 11th to the 25th, very numerous and immensely large. On the 17th, the measurement of one was 84,000 miles in length, and over 27,000 miles in width. On the 18th, it measured 78,500 miles in length by 41,000 miles in width; on the 21st it divided and broke up into groups.

EARTHQUAKES.—On the 7th a violent earthquake in the East

Indies. From 9 A. M. the volcano Ternate gave out a dull, rumbling sound, with loud reports at intervals, continuing through the night and all next day, with streams of lava. The sky was black and the whole landscape darkened with smoke. At day break on the 8th, the outburst of lava was so great that the inhabitants began to fly to the neighboring islands. The eruption of fire and lava and stones continued twelve days, when activity somewhat abated. After a short interval of comparative rest, another terrific explosion took place, which leveled many buildings to their foundations. The whole island reeled, and the damage to plantations and houses was enormous. Aug. 20th, a violent earthquake accompanied by a loud rumbling noise in Jamaica. 21st, a severe earthquake felt at Callao, Peru, at 8.30 A. M. Undulations from Northwest to Southeast. The same earthquake was felt at Cero Azul, Pisco, etc. At the time of the shock the sea was calm, but it suddenly became rough, and a strong Southeast gale set in. A ship, 200 miles from the coast, felt the shock at the same hour. The sea instantly became agitated and remained disturbed for two days.

MAGNETIC DISTURBANCES.—Aug. 16th and 17th, violent magnetic disturbances observed in Cuba and at several observatories in Europe. Those at the observatory at Havana, very violent and remarkable. 21st, extraordinary variations in the magnetic needle observed at the observatory at Havana, Cuba. 24th, another powerful magnetic disturbance observed, of several hours duration.

REMARKS.—By comparison of dates it will be observed that the violent magnetic disturbances of the 16th and 17th were synchronous with the large sunspot, with the brilliant auroras of the 16th and 17th, and with the Florida cyclone; and those of the 21st with the breaking up of the large sunspots, the fiery aurora, the earthquake in Peru, with the terrible cyclone and earthquake at St. Thomas. The magnetic disturbances of the 24th were synchronous with the continuous aurora from the 18th, with the tropical hurricane on its way from St. Thomas to the coast of Florida; with the terrible typhoon at Yokohama, and terrific tornadoes in various parts of the country.

The coincidences in time of so many and so different and various phenomena, make the inference unavoidable that they are corelated and originate in one and the same cause, which we have elsewhere shown to be Electricity.

CYCLONES, ETC.—August 12th to 14th, terrible rainstorms in the southern parts of England and Wales, doing immense damage. 16th, a severe and destructive cyclone in the interior and along the coast of Florida, continuing to the 18th, eleven and a half inches of water fell during the time. 21st, a destructive water-spout burst over the village Ollon, Switzerland. 20th, St. Kitts devastated by a hurricane. 21st, the same hurricane swept over St. Thomas. Every house was destroyed, and the whole place laid in ruins. At 1 A. M. the gale was East; shortly after, North-East, blowing furiously at noon; then it veered to the North, when a terrific hurricane fell upon the Island, shifting to the North-West; it blew with great violence to 5 P. M., when a lull occurred which lasted one hour, when a terrible hurricane again broke out, this time from the South, but it did not last long. It was followed by a terrible gale from the South-East till long after midnight. The duration of its greatest violence was two hours.

In the afternoon, during the hurricane, several severe shocks of an earthquake were felt, rendering the situation of the people more terrible, who, amid the crashing of the roofs overhead by the hurricane, felt at the same time the earth beneath their feet reeling from the throes of an earthquake. Upwards of 150 of the inhabitants were killed, and upwards of 6,000 rendered homeless and totally destitute.

On the 25th this same hurricane reached the Bahama Isles, still accompanied by the earthquake. On reaching the coast of Florida, which it did the same day, it swung around to the North-East, visiting successively Georgia and South Carolina in the wake of the destructive hurricane that had passed along the coast on the 20th and 21st. 24th, a destructive tornado at Cristine, Ohio, accompanied with hail and unprecedented torrents of rain. At Yokohama a terrible typhoon raged, swamping twenty vessels loaded with tea, all of which was totally lost. The coal sheds of the Pacific Mail Company were destroyed. The shipping in all the Japanese ports sustained immense damage. A United States store ship lost four boats; three seamen on board were killed and many wounded. August 22d, at Ihangard,

India, amid a terrible storm and an unprecedented downpour of rain, a terrific thunderbolt fell; the earth where it fell was literally burst open, and all the huts together with their inmates were swallowed up in the chasm. More than 60 people perished.*
30th, tremendous heavy rains in Pennsylvania, North to New York and the New England States, causing destructive floods.

On my Journal I find the following under 24th: "The cirrus clouds here are, this afternoon, moving West, which is contrary to the wind. This I take as an indication of an unusual violent cyclone near or on the east coast of either Georgia or Florida." On the 25th is this memorandum: "Since the 20th there has been a fiery lurid haze covering the Valley of the Mississippi from the Gulf north to St. Paul. Does this indicate earthquakes again as it did last June? On the Pacific coast a lurid haze is considered as a prelude to earthquakes." By reference to the records of earthquakes during the month, it will be seen that these five days cover the earthquakes, in Peru, South Pacific Ocean, Jamaica, and that accompanying the great cyclone from St. Thomas to the Bahamas.

AURORAS.—September of this year was unusually noted for the frequency and brilliancy of its auroras. They are recorded on the 3d, very bright,—4th, 5th, 6th, 7th, very brilliant,—8th, 9th, 18th, 19th, bright and fiery red,—20th, 25th, and 30th, bright with orange streamers.

REMARKS.—By reference to records of this month it will be perceived that the most remarkable auroras were again synchronous with the most remarkable phenomena of the month: namely, the very bright one of the 3d with the earthquake in Jamaica, the extremely brilliant one on the 7th, with the solar outburst; the bright and fiery red of the 19th, with the earthquake at Tortola; and the bright with orange streamers of the 25th with the tornadoes of Indiana and North Carolina.

CYCLONES.—Sept. 3d—A destructive typhoon in the Chinese Sea. On the same day a terrible hurricane was raging on the Carribean Sea. It reached the coast of Florida and Georgia on the 5th and 6th. 5th, a terrific tornado and destructive hail-

---

*NOTE.—This was one of those mine-like explosive earthquakes that sometimes but rarely occur; not only when a thunder shower is overhead, but when it is only cloudy.

storm in the eastern part of Nebraska. The tornado lifted and carried away loaded cars from the railroad, and destroyed many houses. 8th, another violent tornado visited the coast of Florida, and on the same day a destructive gale prevailed on Lake Erie, continuing through the 9th and 10th. The tropical hurricane moved down the Atlantic coast during the same time. 25th, a terrific hailstorm at Purcell Station, Indiana, breaking the headlights, smashing the window of the cab and passenger cars, and stopping the train. Same day a tremendous rainstorm and tornado at Raleigh, North Carolina.

EARTHQUAKES.—Sept. 3d, an earthquake at Kingston, Jamaica, while the hurricane raged. Sept. 6th, a tremendous volcanic eruption of Maunaloa, Sandwich Islands. 9th, a severe earthquake at Burgundy, France; many buildings were injured, and stone fences leveled even with the ground. 20th, a telegram from Kingston, Jamaica, of this date, says: "Yesterday the cable to St. Kitts was injured by an earthquake, which rendered 7,000 persons houseless in the island of Tortola." 30th, severe earthquake at Bombay.

SOLAR.—An extraordinary solar outburst was observed by Prof. C. A. Young, at Dartmouth College, New Hampshire.

PHENOMENA OF OCTOBER.—Auroras were observed on the 3d, 4th, 5th, 8th, 9th, 12th, 13th, 15th and 25th of this month.

REMARKS.—The aurora of the 4th was quite bright with an orange tinge and white streamers. It was synchronous with the appearance of a seismic paroxysm that first manifested itself in South America, and lasted seven days, in which time it pervaded the whole world. Chili, Arequipa, Peru generally, were shaken. On the 8th and 9th, the island of Mindanao, East Indies, was shaken, and the earth rent by an earthquake. Constantinople and New Jersey were shaken on the same day. Auroras were continuous during these seven days, and seen every night when the sky was not overcast with clouds. That of the 9th was very bright and persistent during the whole night. That on the night of the 12th was the most brilliant of the month. I saw it two hours after sunrise on the 13th, and pointed it out to people at the railroad station. The streamers were very distinct and all converged to a point in the magnetic north, about 15° above the horizon. The rays appeared exactly like those of the Sun radiated from behind a cloud near sunset. It was synchronous with the sand cyclones of California and Nevada, the snow storm on the mountains that blocked up the Union Pacific Railroad, the violent hurricane in Nova Scotia, and the tremendous rainfalls that caused such destructive freshets in the New England States. It was also the precursor of the heavy rain storm approaching from the northwest, which reached St. Louis on the night of the 13th, and that did so much harm, for several days after, on the Lakes and in the valley of the St. Lawrence.

EARTHQUAKES.—On the 4th of October a seismic paroxysm manifested itself at Arequipa, Peru. It spread itself like a wave and in seven days covered the whole world. It was felt severely at Chiriqui on the 4th and 5th, and violently on the 5th at Iquiqui. The towns of Rica and Matilla were totally destroyed on the 6th. Many of the adjacent towns suffered severely in buildings and other property, and in the loss of human lives. It was felt at sea by a steamer from Panama to Callao. On the 8th, severe shocks were felt at Constantinople, and two shocks, twelve hours apart, were felt in Delaware and New Jersey. A terrific earthquake occurred at Pollok, Mindinao, on the 8th and 9th, rending the earth and giving rise to sulphurous springs. 15th, shocks of an earthquake felt in the New England States. It was very strong at Concord, New Hampshire. 18th, an earthquake felt at Augusta, Maine, at 4.40 P. M,; it was very strong, and its duration from ten to twelve minutes. 25th, a destructive earthquake at Bajo, Chili.

CYCLONES, ETC.—Oct. 7th and 8th, violent typhoons raged at Shanghai and other ports on the Chinese Sea. The loss of ships and property destroyed was immense. 12th, a terrible sand cyclone in California, between the Colorado and Mohave rivers. Near Fort Tijon, it was estimated that nearly 50,000 sheep were destroyed by the sand storm. On the same day sand storms also occurred in Nevada. Oct. 12th and 13th, a violent snow storm raged on the Union Pacific Railroad where it crosses the Rocky Mountains, blockading the road for seven days between Rawlings and Sherman stations. At 6 P.M. the heaviest thunder shower for years struck St. Louis. It continued all night, and immense quantities of rain fell. 13th, a violent hurricane occurred at Halifax, destroying an immense amount of property on shore, and doing great damage to the shipping in the harbor. On the same day heavy rainfalls in the New England States, especially in Maine, where they produced disastrous floods. 14th, heavy gales on the Lakes; many schooners wrecked and lives lost. Terrific gale and rainstorm at Montreal, all day and night. Many frame houses blown away; the brick walls of the gas building blown down; great damage to property generally; and several lives lost. 16th, another destructive hurricane at Halifax. This was a continuation of the storm that was passing over the Continent from the West for the last four or five days.

An examination of the record of these eighty days, leaves no doubt of the character and species of the phenomena that occur at a physical disturbance. Nominally one of these phenomena is a variation in the intensity and in the direction of force of telluric Magnetism; but really it is a variation of intensity and a change of direction in the electric currents circulating through the Earth, as we show at the proper place. The other phenomena are, sunspots, auroras, earthquakes, cyclones, and heavy rain-falls. Whenever special observation is once made, it will be ascertained, that frequent, sudden, and extreme oscillations of atmospheric pressure, is also one of the accompanying phenomena.

Since it is a cardinal article in our creed, that no hypothesis is legitimate that is not susceptible of verification, therefore we never permit any such hypothesis to vex us. To be treated with respect and attention, a hypothesis must either show in its character or on its face, that it can be either verified or disproved by facts. When a number of facts seem to make the hypothesis plausible, or warrant us in doing so, we elevate it to the rank of a theory; but of every theory we require before its acceptance as a truth, that it must not only be verified by all the proper facts, but that in turn it must explain them; yea, more, we exact of it that it must suggest facts, not known before, nor even suspected.

Our theory, it will be perceived, is, that planetary equinoxes are the causes of the disturbance to which our Earth and its Atmosphere is periodically subject. The great Jovial Cycle has been long known; not however as the Jovial, but as the Eleven years Cycle. It is hence but a question of fact whether or not the length of the Cycle corresponds, as we contend, with the

Jovial year. Facts alone must be the arbiters to decide this point, for they alone are competent, and their decision must be accepted as conclusive and final.

We have seen the species and character of the phenomena that occur and prevail at the critical period of the Cycle. We see that some of them are of such an imposing character that they not only must have attracted, but compelled attention from the very Infancy of the Human Race. Hence, in the historical period, they must have passed upon the record. For the verification of the theory, therefore, we must appeal to history, to hear what it has to say or knows about the phenomena. But looking at their character, it is obvious that not all of the phenomena could be observed. Some, because they took place in the Sun, ninety-five millions of miles away, and there were no instruments to aid the natural eye; and others were unknown, and if known, to observe them required more facilities, skill and intelligence than Man possessed. Even of those that were so obvious that they could not escape observation, only a few, the most imposing, would be deemed worthy a place in the record. Of course the record is meagre, but it is ample to satisfy any reasonable person that the great Physical Cycle has the same length as the Jovial year.

Cyclones are amongst the most imposing of physical phenomena. They however do not visit all parts of the Globe with the same frequency, nor with equal grandeur and energy. In fact, in some parts of the World they are not known as phenomena, to be feared and dreaded. Even where they are both frequent and energetic, each one taken singly is confined within so narrow limits, and its devastation restricted to such narrow strip of territory, and, moreover, all traces of its path being so soon obliterated by Man or Time, that the old chroniclers did not deem them worthy of a prominent place in history; and where they mention them, generally the date is so indefinite as to be unavailable for scientific research. But it is not so with earthquakes. They raise and sink islands and continents; heave up the mountains; and from some of them belch forth fire and smoke, and pour out a flood of fiery lava which overwhelms cities. In fact wherever they prevail with any degree of energy, they more or less permanently change the physical aspect and outlines of shores and continents. Hence history at all periods is crowded with their record; not so precise as desirable, nor as exact as they should be when they are to be used for ascertaining and proving their source in, and corelation to cosmical causes that supervene at fixed astronomical periods. The ancient chroniclers did not record them for scientific purpose, but as sensational on account of their astounding character. Hence they never mention the day nor the month, and are even careless about the year. This however is not the case with physical phenomena alone, but is equally true of the records of political events: for by a comparison of dates given for the same event by different authors, either from their carelessness or from errors in transcribing, there is often a difference of a year or two, and sometimes even more. I give the dates as I find them, and where there are discrepancies, I give that considered the best authority. I found by examination however that my dates, with only one or two exceptions, correspond closely with those of Hayden's Dictionary of Dates.

As the fullest record of physical phenomena is that of earthquakes, so for the purpose of verifying our theory, we will compare the time of their occurrence with that of the Jovial Cycle of 1871. But before doing so, we will remind the reader of three facts already stated, namely, (1) The Jovial disturbance by actual observation has been ascertained to extend over a period nearly, if not quite three years. (2) That though, during the period of perturbation the incessant occurrences of the characteristic phenomena in mild and modified forms, show a general excitement of the Earth and the Atmosphere, yet violent paroxysms hardly ever take place from a pure Jovial cause; and, (3) That the violent paroxysms are sequences of a Martial, Venusian, Terrestrial or Mercurial equinoctial disturbance being superimposed on the Jovial. Hence too, these causes frequently accelerate or retard the maximum paroxysm. To this fact I will add that, a Saturnian disturbance frequently prolongs the Jovial perturbation, as well as intensifies those of other planets, as will be shown hereafter. A Saturnian perturbation once in a great while—for obvious reasons—during seven or eight cycles of fifty-nine years, will intensify at the critical period every fifth Jovial disturbance. This cycle has been distinctly noted and determined by observers, and called by them the Fifty-nine

Years Cycle; but they never suspected the cause, namely, that five Jovial years are only one hundred and ten days more, and two Saturnian years only thirty-three days less than fifty-nine years.

It is hence perceived, that every second revolution of Saturn approaches or recedes 143 days from every fifth of Jupiter; and that it will take as many times 59 years as 143 is contained times in the length of the Jovial perturbation—usually estimated three years—before the Fifty-nine Years Cycle will disappear, to reappear after a long interval. As all planets have two equinoxes in their year, of course there is a double cycle of fifty-nine years. The minor cycle or approximate conjunction of a Jovial and Saturnian equinox took place in 1818, and hence one will occur in 1877. A Saturnian disturbance endures fully for six years. After a careful investigation of the facts, I feel warranted to infer that approximately the equinoctial disturbance of every planet is felt for nearly one-fourth of the period of its revolution around the Sun; generally two-fifths of the time before, and three-fifths of the time after the occurrence of the equinox; except in the case of Jupiter and of the Earth where these proportions generally seem reversed. The following appear to be the length of each planet's perturbation:

|         |            |
|---------|------------|
| Vulcan, | Not known. |
| Mercury,| 15 days.   |
| Venus,  | 56 days.   |
| Earth,  | 3 months.  |
| Mars,   | 5 months.  |
| Jupiter,| 3 years.   |
| Saturn, | 7 years.   |

Vulcan's equinoxes probably occur near his nodes, approximately 4° and 184°. Judging from his appearance while on the disc of the Sun in September, 1859, his size must be fully equal to, if not much greater than, that of Venus.

### VERIFICATION OF THE CYCLE.

EARTHQUAKES.—As already stated, earthquakes are not only the earliest but almost exclusively the only phenomena recorded in primitive history. We hence avail ourselves of historical earthquakes to test the truth of our theory. But in order to show that our sole object is to ascertain the truth, we have taken the lists, prepared by others, of remarkable earthquakes, and not a select list of our own. We have only added one or two to the list, such as had evidently been overlooked by the compilers.

Smith, in his valuable History of Greece, speaking of the Lacedæmonians, says: "In the year B. C. 464, their capital (Sparta) was visited by an earthquake which laid it in ruins, and killed 20,000 of its citizens, besides a large body of their chosen youth, who were in a building at their gymnasium exercises." Other dates are given for the happening of this event, as the years 465 and 466. According to the date given by Smith —as only the year is given—the earthquake was about synchronous with the great cyclical disturbance of Jupiter. From A.D. 1871.72, when the last Jovial disturbance occurred, to B. C. 464 are 2350.72 years; divided by 11.86 years, the length of the Jovial year, gives 197 revolutions of Jupiter, lacking seven-tenths of a year. The Jovial equinox hence occur in the year B. C. 465, and the earthquake must have happened 8 or 10 months after the equinox. If the earthquake occurred as stated in the year 465, then it was synchronous with the equinox; if, in the year 466 B. C., then it happened before the equinox. All these dates bring the earthquake within the limits of a Jovial disturbance. For convenience sake we will take the year 1871, discarding the fraction, unless otherwise stated, as our standard. We do so for convenience sake, and for the reason that we have neither day nor month given in the year for the events recorded in history. The year B. C. 425 is given for the earthquake that made Eubeca an island. From our standard to B.C. 425 both inclusive, are 2296 years, or 193½ Jovial Cycles, plus one year. If this event is not apocryphal, the earthquake occurred some ten or twelve months before the minor* equinox of Jupiter. Ellice and Bula, two cities in Peloponessus, were swallowed up in the year B. C. 372. The time elapsed from the year B. C. 372 to our standard was 2243 years, equals 186 Jovial cycles, plus 17 months, that is, the earthquake occurred from 10 to 17 months before the major equinox.

---

*Note.—Half a cycle shows that it was not the major but the minor disturbing equinox that occurred. We designate the equinoxes accordingly.

The earthquake at Rome, according to Livy, into whose chasm, armed and mounted on a stately horse, M. Curtius leaped, occurred B. C. 358. From 1871.72 to 358 are 2229.72 years, equal 188 Jovial Cycles, plus about 36 days. This earthquake hence was within 36 days of the major equinox. Durcas, in Greece, and twelve cities in Campania, were engulfed with all their inhabitants by an earthquake in the year B. C. 345, or 187 cycles, plus 1.1 years from 1871. Lysimachia and its inhabitants was totally buried by an earthquake B.C. 283, or from 1871 the time elapsed was 181½ Jovial cycles, plus 17 months, that is, the earthquake occurred at the minor equinox, within seventeen months before it.

An earthquake destroyed Ephesus, and other cities in Asia Minor, A D. 17. The time from 1871 to 17 is 1854 years, or 156½ cycles, minus one year. The earthquake therefore occurred within one year of the minor equinox.

In the year A. D. 79 occurred that extraordinary earthquake and eruption of Vesuvius,—up to that time supposed to be an extinct volcano,—that overwhelmed the cities of Herculaneum and Pompeii. This is one of the most appalling phenomena recorded in history. It happened on the night of the 24th of August, 79, when many of the people were at the theatres. From the 24th of August, A.D. 79, to 25th of September, 1871, are 1792.08 years, equal to 151 Jovial Cycles, plus 1.14 years. The equinox occurred October 14th, A.D. 80, or nearly 14 months after the earthquake. It is proper to remark here that a Saturnian equinox occurred only a few weeks after the Jovial in the year 80. Hence the paroxysm was not only accelerated but intensified by the near conjunction of the two equinoxes.

The eruption and earthquake are the most terrible and appaling that history furnishes. The cities of Herculaneum, Pompeii and Stabia, were overwhelmed with lava and buried in ashes, after having sustained a total overthrow by the earthquake; and the surrounding country was deeply covered with scoriae. Over 250,000 persons perished; the elder Pliny lost his life, in the interest both of Science and Humanity.

We may as well here consider all the subsequent seismic phenomena of Vesuvius, as they afford very pointed confirmation of our theory. But before doing so, we will point to a historical fact that a portion of Pompeii was destroyed in A.D. 63, exactly at the minor equinox of Jupiter, that is one and a half cycle before the great catastrophe.

One of the most destructive eruptions of Vesuvius occurred on the 24th of November, 1759. Jupiter passed his major equinox in January, 1759, and Saturn in November, coinciding almost, if not quite, with the eruption. Besides this, Venus passed her equinox October 29, or only 26 days before the eruption.

In June, 1794—day not given—one of the most destructive eruptions occurred. The lava flowed over 5,000 acres of vineyards and cultivated lands. The town of Torre del Greco was burnt; and the top of the mountain fell in, so that the crater is now nearly two miles in circumference. In 1794 the minor equinox of Jupiter occurred in August, and therefore coincides with the eruption, within two months. The violent eruption of May, 1855, occurred one year and four months after the minor equinox of Jupiter, and within 22 days of a Venusian equinox. There were a series of violent eruptions that commenced in May, 1858, and continued at intervals to December, 1861. Jupiter's major equinox occurred in December, 1859, and a Saturnian early in the year 1863. This was the cause of the prolongation not only of the Vesuvian eruptions, but of violent earthquakes and electric disturbances from the beginning of the year 1858, till after the minor equinox of Jupiter in October, 1865, involving all parts of the Globe. From 1858 to 1861 all the phenomena fell within the disturbed period—three years—of Jupiter.

As the phenomena and their relation to the Jovial equinoxes, are now before the reader, we may as well show their more intimate relations to those of Venus. We therefore give the following facts as illustrations of how a crisis is brought on during the prevalence of a Jovial or Saturnian perturbation by the superimposition of the disturbance occasioned by an equinox of one of the inferior planets. The destructive eruption of 1794 occurred in June; an equinox of Venus took place on the 6th of June of that year. The eruption of 1855 occurred in May and June; an equinox of Venus June 25th. The eruptions of 1858 in May and June. Equinoxes of Venus in 1858 are one in April and one in July. The destructive eruption of 1859 occurred in

June. An equinox of Venus occurred June 26th, 1859. For 1860 our information is vague. The record before me only says, "great destruction was caused by the eruptions in the spring and summer of 1860." The spring equinox of Venus occurred in February, and the summer one about the first of June in that year. The violent eruption of December, 1861, which again destroyed Torre del Greco, was synchronous with a Venusian equinox that took place December 9th. And the eruption of 1865 was only 26 days after the Venusian equinox of January 10th.

Returning now to our list of earthquakes, which we had followed up the year A. D. 79; and found that the engulphing of Pompeii and Herculaneum occurred 14 months before the major equinox of Jupiter, which did not take place until the following year, say A. D. 80.86. The next earthquake on the list is indefinite, which reads as follows: Four cities in Asia, two in Greece, and two in Galatia overturned by an earthquake in the year A. D. 107 Two cycles of Jupiter, namely, 23.72 years from the year 80.86, were completed 2.4 years before A. D. 107; and one cycle of Saturn 3.3 years after 107. In all these approximate periodical conjunctions of Jupiter and Saturn, the minimum of perturbation between the major and minor equinoxes of Jupiter is obliterated, and the perturbation is continuous for seven or eight years; and this may account for the happening of this earthquake so remote from the cyclical period of both planets as not to be referable to either.

In the year 115 Antioch was destroyed by an earthquake. This occurred one year and five months before a major equinox of Jupiter. Nicomedia, Cæsarea and Nicea were overturned by earthquake in 126. This occurred within two months of a major equinox.

In 357, in Asia, Pontus and Macedonia, 150 cities were partly destroyed and damaged by an earthquake. This occurred 17 months before a minor equinox.

Nicodemia totally demolished, and its inhabitants buried beneath its ruins, in the year 358. This accords within a few months with a minor equinox.

In 543, an earthquake was felt in all parts of the world. The year exactly coincides with a major equinox.

In 557, a great many edifices destroyed, and thousands perished by an earthquake at Constantinople. This seems too remote from either a Jovial or Saturnian perturbation to be referable to either. If it be assumed to have occurred at the beginning of the year, it was twenty months before a major equinox.

In Africa, in the year 560, many cities were overthrown by an earthquake. This occurred about five months after a major equinox.

In the year 626, Antioch, in Syria, was destroyed by an earthquake. Over 250,000 persons perished in this calamity. This occurrence took place within a few months of a major equinox.

In the year 742 an awful earthquake occurred in Syria, Palestine and Asia; more than 500 towns were destroyed, and the loss of life surpassed all calculation. As this event happened when Jupiter was about midway between his equinoctial points, it cannot be referred to him; but as it occurred within 17 months of a Saturnian equinox, it was partly owing to a Saturnian disturbance brought about as usually by an equinox of either Mars, the Earth, or Venus. Mercury may have contributed its influence, which however lasts only a few days; besides, either Uranus or Neptune may have been at their equinoctial points, and thus aided in producing the convulsion.

In the year 801, all Europe was shaken by an earthquake. This again, for the reasons above stated, is a purely Saturnian phenomena; it occurring 18 months before his equinox.

A severe earthquake was felt in England, and throughout Europe, in 1089. There are exactly 66 Jovial cycles from 1089 to 1871-72.

In the year 1114 another dreadful earthquake occurred at Antioch. Many cities were destroyed in Syria, among them Mariseum and Mamistria. This earthquake occurred one year and three months before a major equinox.

The earthquake of 1137 in Sicily, by which Catania was overturned and 15,000 persons perished, occurred within two months of a major equinox.

The severe earthquake in England in 1142, occurred within a few months of a minor equinox.

The earthquake in Calabria, in which one of its cities was thrown into, and engulfed by, the Adriatic Sea, occurred in

1186. This took place over two years after an equinox of Jupiter, it cannot therefore be referred to him, but it occurred within six months of a Saturnian equinox. As the day and month are not given, we can only give the extreme limits, namely, six months.

The earthquake in England, by which Glastonbury was destroyed, occurred in 1274, and was one year and two months anterior to a minor equinox.

The severest earthquake ever known in England, occurred on Nov. 14th, 1318. It took place just one year and three months before a minor equinox.

In 1456, an earthquake occurred at Naples, in which 40,000 persons perished. This earthquake corresponds exactly with the year of the major equinox.

The earthquake of February, 1531, by which 1400 houses in Lisbon were destroyed, and 30,000 persons in that and neighboring cities buried, does not coincide with either a Jovial or Saturnian cycle, being two and three-fourths years removed from the first, and about six years from the last. A Venusian equinox occurred on the 18th of January of that year, and a Martial in February, and they may have caused it, but this is a rare occurrence.

On the 19th of Sept., 1538, Monte Nuovo, (New Mountain,) was raised by an earthquake near Pozzuola, Italy. This occurred not quite a year before the major equinox.

In 1580, part of St. Pauls and the Temple churches fell from the effects of an earthquake. This occurred about five or six months before a minor equinox.

The year 1596 is given for a severe earthquake in Japan, in which several cities were made ruins. This date does nor correspond with a Jovial cycle, but with a Saturnian one within one and a half years.

A destructive earthquake in Calabria, in 1638. If it occurred in the latter part of the year, its occurrence was within about eight months of a minor equinox.

On the 11th of June, 1638, a new island was thrown up by an earthquake near St. Michaels, Azores. In December, 1719, the island having disappeared, arose again above the surface of the sea. This was during a Jovial and Saturnian disturbance. On the 13th of June, 1811, it arose a third time, but the village of St. Michaels was sunk the year before, as will appear at the proper place.

An earthquake occurred in China, in 1692, by which 300,000 persons perished in Pekin alone. This earthquake could not have occurred more than 18 months after a minor equinox; and it may have occurred within six.

The earthquake by which Port Royal, Jamaica, was sunk 40 fathoms beneath the sea, occurred in 1692. This event took place about one year and one month before the major equinox.

In the year 1693, an earthquake overturned 54 cities and towns, 300 villages in Sicily. Of Catania, with its 18,000 people, not a vestige remained; over 100,000 lives were lost. As the earthquake occurred in September, we find it occurred within 36 days of the major equinox.

This brings the record down to the beginning of the Eighteenth Century, and we will now take a retrospect to see what the testimony of the facts so far has been, and judge whether they prove or disprove the theory that the length of the Cycle of Physical Perturbation corresponds with the Jovial year; and that the Jovial equinoxes are the disturbing causes. It is admitted that the major perturbation, which we have endeavored to identify with the first equinox of Jupiter after he has been to his perihelion, and where he is 47,000,000 of miles nearer the Sun than at aphelion, far exceeds the energy of the minor perturbation after passing aphelion. The inference hence is that, not only the larger number of phenomena, but those of greatest energy, must occur at what we call the period of the major equinox. But before we investigate this point, let us take a general survey of the ground.

From the year B. C. 465, to A. D. 1700, are 2165 years. Since a Jovial equinox occurs every 5.93 years, therefore in 2165 years 365 such equinoxes occur. Jupiter's year is within a small fraction of 12 years. Actual observation has determined that ordinarily the major perturbation lasts about three years, from its first feeble manifestations until it disappears; the minor is several months shorter. We can hence assume that in the twenty-one and two-thirds centuries the Earth for nearly one-half the time, say 1,000 years, was under the Jovial influence, leaving 1165 years when no such influence was perceptible. We now divide the thousand years into 365 parts—the number of Jovial equinoxes in the twenty-one and two-thirds centuries—and disseminate these parts equally amongst the 2165 years. Our theory assumes that in the 2165 years there are 365 periodical disturbances; and it exacts that each of these disturbances shall fall upon one of these distributed periods; and that unless each does so, the theory falls to the ground. In this long period of 2165 years, not more than one-fifth of the Globe was under observation. Of the hundreds of phenomena observed, not a tithe were deemed worthy to be recorded. In fact only the most extraordinary and appalling ones were deemed worthy of a place upon the records. At the end of the period, a virtuoso in search of the curious and marvellous, finds forty-four of the phenomena that history for 21 centuries has collected and recorded, so extraordinary that he puts them upon his "List of the most dreadful and appalling earthquakes in all ages."

Well, whatever scientific value facts that were discarded from this list may have possessed, these extraordinary ones possess in a more eminent degree. These latter we now take up singly, and compare their dates with the fixed periods the theory assigned in over twenty-one hundred years. This certainly will severely test the theory; but behold the result! Of the forty-four, thirty-six fall upon the periods fixed by the theory; four fall upon a Saturnian period, deduced from the assumption of the truth of the Jovial period; one falls upon a period formed by an accidental conjunction of a Martial and a Venusian cycle, whose existence were also inferred from the Jovial cycle; one is the joint product of the Jovial and Saturnian periods at an approximate conjunction; and to but two of the phenomena no periods can yet be assigned; and there is even a possibility that their dates may be erroneous.

Returning now to the verification of our two inferences, we find upon review the following facts; Of the thirty-six phenomena occurring at the Jovial periods, twenty-two occur at the major and but fourteen at the minor equinox. The inference therefore is confirmed that the large portion of the phenomena would occur at the major equinox. The second inference is that the phenomena occurring at the major equinox would upon examination prove to have greater intensity and energy than those at the minor, is also confirmed; for eleven of those that occurred at the major equinox exhibited intense energy, while but two of those at the minor were so characterized.

So far the facts seem to indicate that Saturnian phenomena nearly uniformly manifest extraordinary energy and violence.

During the Eighteenth and up the present time of the Nineteenth Century, the records of earthquakes are too numerous to be considered singly, we will therefore only select a few of the most extraordinary ones of modern times, and then proceed to discuss another point of the subject

In 1731, more than 100,000 people lost their lives by an earthquake in Pekin, China. This was a year and a half after a major equinox.

In 1736, a year after the minor equinox, a mountain in Hungary was turned around by an earthquake. This was one of those rare phenomena, a rotary earthquake, of which that at Riobamba was an instance. Humboldt says of the Riobamba earthquake: "I was shown a place where the whole furniture of one house had been found under the ruins of another: the earth evidently had moved like a fluid in streams or circular currents, the direction of which was first downwards, then horizontal, and lastly upwards." The great earthquakes of Lisbon and Calabria, in 1783, were conspicuous examples of these rotary earthquakes. In the latter, by the whirling motion of the earth, buildings, without being overturned, were twisted in different directions, parallel rows of trees were deflected, and in fields producing two different kinds of grain, one crop was made to take the place of that occupied by another. This displacement of lands gave rise to much litigation.

In 1740 a violent earthquake occurred in Asia Minor and Southern Europe. At Palermo a convent fell into a chasm and was swallowed up. This occurred the year before the major equinox.

Lima and Callao, Peru, demolished by an earthquake October 28th, 1746. This occurred not quite five months before a minor equinox.

Adrianople nearly overwhelmed by an earthquake 1752; about six months before a major equinox.

Grand Cairo destroyed by an earthquake. More than half the houses swallowed up, and 18,000 or 20,000 lives lost. This took place in 1754, the year after the major equinox.

As there was an approximate conjunction of the next Jovial equinox which took place 1759.09, and a Saturnian taking place in 1759.90, the usual prolonged perturbation in such cases, about eight years, took place, extending over the intervening period, and including the previous major Jovial equinox, which occurred in 1753.16. The earthquakes were unusually frequent, and of the most violent character during this period. In April, 1755, the city of Quito was swallowed up. On November 1st, 1755, occurred the great earthquake of Lisbon. In eight minutes nearly all the houses and 60,000 of the inhabitants were swallowed up; whole streets were buried. The cities of Coimbra, Oporta and Braga suffered dreadfully; St. Ubas was totally overturned. A large part of Malaga was in ruins. One half of Fez, in Morocco, was destroyed, and more than 12,000 Arabs perished. One-half of the island of Madeira was made waste; and 2,000 houses in the island of Mytelene were overthrown. The shock was felt in North and South America, in Scotland, Norway, and eastward into Asia. A Venusian equinox occurred on the very day that this earthquake occurred. In 1759, at the approximate conjunction of the equinoxes of Jupiter and Saturn terrible earthquakes raged over North and South America, Southern Europe, Syria, Asia and the East Indies. The city of Balbec, in Syria, was totally destroyed.

On the 5th of February, 1783, not quite four months after the minor equinox, Messina, and other towns in Sicily, were overthrown, and over 40,000 persons perished; and in Iceland, January 9th, 1783, a whole river and its valley was engulfed by an earthquake.

At the major equinox of September 29th, 1788, there was a general earthquake commotion throughout the Globe. The most noted was that of September 30th, 1789, just one year after the equinox in which Borgo di San Sepolcro was swallowed up.

At the major equinox, August, 1800, amongst others, was the historical earthquake at Constantinople, which laid a great portion of the city in ruins, amongst them the royal palace. It occurred on the 26th of September, just one month after the equinox.

In 1804, another Saturnian, and in 1806, the Jovial minor equinox occurred. In 1804, the most violent earthquake ever known was felt in Holland, and on July 26th, 1805, at Frosolone, and at Naples 6,000 to 8,000 lives were lost.

The major equinox of May, 1812, is memorable for its intense, violent and general seismic convulsions. About twenty months before the equinox the village of St. Michael, Azores, was sunk, and a lake of boiling water appeared in its place. In March, 1812, two months before the equinox, the city of Caracas, (as given at another place) was destroyed. But the most memorable of all the attending earthquakes of this equinox, was that of the valley of the Mississippi, commencing Dec. 16th, 1811, and continuing through 1812 and 1813. In this earthquake several islands in the Mississippi, near New Madrid, were sunk, and the river at one time driven back eighteen miles, overflowing the adjacent country. Half of the County as well as the village of New Madrid, were submerged. Several new lakes were formed, one sixty miles long and several miles wide. An immense area of forest was sunk below the water level, forming what is now known as the Great Earthquake Swamp. The Earth's surface rose in huge undulations like the billows of the sea, and with terrific detonations, chasms yawned from which vast columns of sand, mud, water, and a substance resembling coke, were ejected. The whole face of the country in that region underwent a permanent physical change.

The minor equinox occurred 1818.39; and as Saturn passes an equinoctial point in every 14¾ years, a Saturnian equinox occurred early in 1819. Earthquakes were general all over the Globe, and many very destructive. The most remarkable one occurred in June, 1819, by which the district of Kutch was sunk, and several thousand people perished. Geneva, Palermo, Rome, and many cities in Southern Europe suffered during 1819; immense damage was done to property, and many lives were lost.

The equinox of 1824.32, shows nothing remarkable; though there were frequent earthquakes in all parts of the world; yet they were of a mild character. The great paroxysm that came on in 1810 and lasted to 1819, in which two Jovial and one Saturnian equinox occurred, seems to have exhausted the frame of Nature, and there was a period of comparative repose until 1829, a short time before the minor equinox, which took place 1830.35.

On March 21st, 1829, Mercia and numerous towns and villages in Spain, were laid waste, and 6,000 persons perished. This was one year before the minor equinox.

Another Saturnian equinox occurred in 1834, and consequently we have another prolonged seismic paroxysm until after the Jovial major equinox in 1836.18.

The following are the most noted earthquakes during this period:

No less than forty shocks were experienced in Italy during 1834. At Pontremoli, Feb. 14th, many houses were demolished and many lives lost—not a chimney was left standing.

Cozenza and many villages destroyed in Calabria, and many lives lost, April 29th, 1835.

Rossano and other villages destroyed, over 1,000 persons perished, Oct. 12, 1836.

Many cities in Southern Syria totally demolished, and thousands of lives lost, in December, 1836.

At the minor equinox of 1842.11, we have: two-thirds of the town of Cape Haytien, St. Domingo, destroyed, and between 4,000 and 5,000 lives lost, May 7th, 1842.

Point a Pitre, Gaudaloupe, utterly destroyed on February 8th, 1843.

The major equinox of 1848.04, was marked by general seismic convulsions all over the globe, but nothing remarkable occurred until the intervention of a Saturnian equinox in the latter part of the year. Then a paroxysm set in, which lasted till after the minor equinox of Jupiter, December 20th, 1853. Some of these earthquakes were most violent and destructive; but they are too many for record here.

In 1859.90, occurred a major equinox, preceded for nearly two years by some of the most violent earthquakes on record. Montemurro, Calabrai, and 22,000 people destroyed Dec. 16th, 1857; Corinth, Feb. 21st, 1859; Quito destroyed and 1,000 persons killed March 21st, 1859; Erzeroum destroyed, and 1,000 killed, June 2d, 1859; San Salvador, Dec. 8th, 1859.

In 1863, another Saturnian equinox intervened with the usual violent paroxysms that continued down till two years after the minor Jovial equinox in 1865.83.

We can only give an abstract of the most prominent. Perugia, Italy, laid waste 1861. Mendoza, South America, destroyed; 7,000 lives lost, 1861. Corinth, and many other cities, Dec. 26, 1861. Guatamala, 150 buildings and fourteen churches destroyed; and many lives lost, Dec. 19, 1862. Thirteen villages destroyed and many lives lost in Asia Minor, April 22d, 1863. Manilla destroyed, the destruction of property immense, and 10,000 persons perished, June 3d, 1863. In Sicily, Macchia, Bendinella, and other villages, destroyed, with great loss of life, July 18th, 1865. The latter belongs to the Jovial equinox which occurred in November of this year.

Part of the phenomena including earthquakes of the major equinox of Sept. 25, 1871, have already been given in the quotation of the general phenomena from Aug. 5, to Oct. 25, 1871. Others will be given when we make quotations for the Venusian Cycle, and therefore need not be repeated here.

The testimony of the facts presented, incontestably establishes these points: that earthquakes have a periodicity in the frequency of their occurrence, and that they show well-defined periods of maxima and minima, which alternate regularly as to time with each other. Examination of the dates of their occurrence shows that these maxima and minima are covariants with those of other physical disturbances; and moreover that the maxima are synchronous with Jupiter's passage through his equinoctial points; the minima in the meanwhile corresponding in time with his passage through the aphelion or perihelion points, situated midway between the equinoctial points. One great point has, therefore, been established in Meteorological Science. But incontestably true as it is, yet it is not at all times owing to one and invariably to the same cause, but oftentimes is the resultant of several causes whose periods for the time being accidentally coincident, the phenomena become often very complicated, and frequently the minimum period is entirely obliterated. Hence we may become bewildered when looking only in one direction for the cause of everything that is taking place under our observation. We must remember that though amongst the ancients Jupiter was regarded as the Earth-shaking deity, as well as the

Heaven-shaking one, or Thunderer, yet we must learn what our facts teach us that the modern Jupiter cannot claim exclusive prerogatives in either Heaven or Earth, for his more sluggish brother, Saturn, once in a while steps in and does his part of the shaking and thundering, vigorously keeping it up for five or six years, and Venus now, as in ancient times, intervenes to aggravate the already fiercely raging turmoil. We have already adverted to the fact that there is an approximate conjunction of the minor equinox of Jupiter and one of Saturn in the year 1877; the former occurring in August, the latter in December. We are therefore on the eve of a seismic period that will last for seven or eight years, and which may bridge over all the time intervening between now and the Jovial equinox in the summer of 1883.

It is therefore probable that frequent earthquakes, many of them of great violence, will occur between now and the close of the Jovial excitement, some time in 1885. This we say in the interest of Science, and not for the purpose of alarm; for, if Science is worth any thing, it must enable us to divine the Future as well as to explain and understand the Past and the Present. Whatever the ignorant or weak-minded may think, and whatever effect the assertion may have upon their feeble minds, there is no cause for alarm whatever. The fears and consequent distress of the good old lady that the boiler would explode while crossing on a horse ferry-boat, were just as reasonable and well grounded as the fears of an earthquake catastrophe in our country. Such an event is an impossibility here, where the necessary conditions for its occurrence do not exist, excepting in some portions of the Pacific States. Earthquakes are caused by disruptive discharges of Electricity through the strata of the Earth. Electric currents, at all times, are circulating through the Earth from East to West. In times of physical perturbations, indicated by sunspots, auroras and great oscillations in the magnetic needle, earth currents, as they are called, often become too intense to be transmitted through the strata of the Earth, unless where the strata are unbroken or of good conductive capacity. These currents where the strata are broken up, or of too feeble conductive capacity, become dammed up, as it were, until they are strong enough to force a passage, which is effected by what is called a disruptive discharge. Hence, often in South America a roll of subterranean thunder is heard before the throes of the earthquake are felt. Now in the Mississippi Valley the strata are unbroken. This is conclusive evidence that they have had the capacity to meet all demands for transmission of Electricity in the Past; and also, that they have the ability to do the same in the Future. This is the reason why no serious earthquake has occurred in our history. The same is true of England and the northern portions of Europe. No earthquake conditions exist there adequate to produce those awful convulsions that so frequently take place in Mexico, South America, Southern Europe and Asia. It is natural that the approach of an earthquake period should be regarded with dread and apprehension in such countries as Campania, near the base of Vesuvius; or Calabria, near the foot of Mount Etna; but in our own country such fears are groundless, and therefore childish.

It is undeniable that a cycle of physical perturbation is always marked not only by frequent but by brilliant auroras. If, upon examination, it be found that the period of the occurrence of a Jovial equinox be similarly characterized, and the periodicity of the auroras to recur, and therefore to coincide with the Jovial equinox, the inference will be unavoidable that the equinox and the auroras must in some way stand to each other in the relation of cause and effect. The ancients not being so much stultified by studying books, were far closer observers of Nature and more assiduous and successful students of her mysteries than we are. The Heavens especially were the object that attracted their close attention, observation and study; because they believed that he who understood them, could read the Future as understandingly as he could the records of the Past. Any extraordinary appearance in the sky was looked upon as a sign and omen that foreshadowed an extraordinary event about to make its appearance. They believed that the destinies of every individual, from the peasant in his hovel to the prince upon his throne, as well as of dynasties and empires, were written by the hand of Fate in the Stars. Hence it was not possible for auroras to have escaped their attention; and consequently could not have failed to find a record either in their legends, poetry or chronicles. Upon examination the facts are found to be as anticipated.

AURORAS.—In the Hindoo Mahabharata, Book I, Chap. 15, (Wilkin's translation) a phenomena is described, in which we recognise, notwithstanding its oriental style, distinctly, the cloud spout, the cyclone, and probably the aurora; though the latter is doubtful, since this phenomenon is rarely seen within the Tropics. It is said to have occurred when the Suras and Asuras were at war, supposed to be about 945 B. C. We quote it to show how early attention was drawn to Cyclones so prevalent in tropical Asia.

"They (the Suras and Asuras) now pull forth the serpent's head [that is the incipient cloud-cone or spout of a tornado] repeatedly; and as often let it go, while there issues from its mouth, thus violently drawn to and fro from the Suras and Asuras, a continual-stream of fire and smoke and wind, which, ascending in thick clouds, replete with lightning, it began to rain down upon the heavenly lands fatigued with their labor."

The reader will recognize in the streams of fire and smoke, the fiery electric cone so often witnessed—of which instances are given in the fore part of this work—in the cloud-spot thrust down upon the Earth, in tornadoes. The reader will also recall that the Prophet Elijah was caught up in a whirlwind of fire. Evidently the fiery cloud-spout of the tornado was known to the Hebrews.

In Hesiod's Theogony there are many passages which show that auroras were well-known phenomena to the ancients. Amongst others occur these lines:

"Through the void spreads a preternatural glare, mingling fire with darkness."

The Chinese records of auroras are the earliest we have; and it is sometimes doubtful whether an aurora is meant or a meteoric shower, we give them as we find them.

Thus we have a French account which says: "In the fiftieth year of the reign of the Emperor Kie or Li-Koue (that is B. C. 1768) the Chinese saw stars falling."

If this was an aurora, then it occurred fifteen months after a major equinox of Jupiter.

The same account says\*: "In the reign of the Emperor Le

---

\*NOTE.—We quote from a translation purporting to give the substance of the Chinese Records.

Wang (B. C. 687) the stars did not appear, and meteors fell like rain." The original statement, however, is: "687 ans avant J. C. les etoiles ne paroissoient pas, il tomba une etoile en forme de pluie." Literally this statement is, in the 687th year before Jesus Christ, the stars did not appear, there fell a star in the form of rain. The latter clause seems to indicate that it was a single bolide that exploded and fell like rain. The fragments of a single bolide certainly, with no degree of propriety, could be compared to a shower of rain, and the statement that the stars had disappeared, makes the translation, that "the meteors fell like rain," inconsistent. We hence infer that the Chinese record, from which the French was taken, meant that there was such a brilliant aurora as to obscure the stars, which dissolved and fell like rain. If this was an aurora then it occurred about 18 months before the major equinox.

In Kaempfer's History of Japan, published in London, 1728, occurs this statement: "A. D. 11—In the 40th year of his (Syn-in's, Emperor of Japan) reign, on a clear and serene day, there arose of a sudden, in China, a violent storm of thunder and lightning; comets, fiery dragons, and uncommon meteors appeared in the air; and it rained fire from Heaven." This certainly is not intended to describe the phenomena of a single day, but of the period when this sudden and violent storm of thunder and lightning, and most probably accompanied by a tornado, arose. The fiery dragons and the rain of fire from Heaven unquestionably mean an aurora; so most probably does the expression "uncommon meteors appeared in the air." The inference hence is almost inevitable that a physical perturbation prevailed at the period. A Jovial major equinox took place the year before.

In the *Chronicum Scotorum* it is recorded that at the time of the battle of Seghais, which occurred A. D. 497, "red blood was brought over the lances," in marching for a night attack. Probably this was the sheen of a fiery aurora on the lances. In the legends of both Scotland and Ireland, as we will see hereafter, the reflection of a red aurora, from the milk they were drinking, the butter they were eating, or of the waters of rivers and lakes, these objects were said to have been turned into blood. From the year 497 to 1871.72 there are 116 Jovial years, lacking one year. The aurora hence occurred the year after the major equinox.

In the same Chronicle where the death by drowning of King Muireertach Mac Erca, in A. D. 531 is recorded, occurs this phrase: "Blood reached girdles on the plain." This has been interpreted to mean that the waters, reflected the light of a fiery aurora. From 531 to 1871.72 are 113 Jovial revolutions, lacking about six months. These events then occurred within six months of a Jovial equinox

In Lynch's *Cambrensis Eversus* are found two records which evidently allude to auroras, but as the dates are not given they are unavailable for our purpose. We quote them to show how indefinite and various the fancies and consequently the descriptions of the chroniclers were. We however are at no loss to know what is meant:

"A. C. 561. Elim Ollfinachta succeeded. He is called Ollfinachta because snow that fell during his reign looked like wine." Under A. C. 673, we have this record: "Finachta succeeded his father to the throne. During his reign an enormous quantity of wine fell like fleeces of snow from the sky."

There is no difficulty in apprehending that the wine that looked like snow, meant the shivering, flickering streamers of a red aurora, which often may be compared to the falling of red or orange red snow.

In the *Chronicum Scotorum*, under A. D. 660, it is recorded: "Darkness at the Kalends of May, at the ninth hour, and the same summer, the sky was seen to burn." The Jovial major equinox occurred in 661, therefore this aurora occurred the year before.

In the same, "A. D. 670, a thin tremulous cloud in form of a rainbow appeared at the fourth watch of the night of the fifth day before Easter Sunday, stretching from East to West in a clear sky; [an auroral arch] and the moon was turned into blood." This aurora was too early by 3.48 years for the major and 2.48 years too late for the minor equinox of Jupiter; that is, Jupiter was then quite near the perihelion, which as far as Jupiter is concerned, is a period of repose.

Upon examination, it is found that a Saturnian equinox occurred 670.21, or within thirty days of this aurora.

"A. D. 680"—*Ibd.*—"Loch n Echnach was turned into blood." The minor equinox of Jupiter occurred in 679.83, or within a few months of the commencement of the year 680, in which the aurora occurred.

In the Anglo Saxon Chronicles, A. D. 685, it is recorded: "In this year it rained blood in Britain, and milk and butter were turned into blood." In September, 685, the major equinox occurred, and hence it and the aurora coincided.

For the better understanding of the obscure annals we are obliged to follow, written by superstitious men, with an undue leaning towards the marvelous, and withal a strong inclination to connect "the portends of the sky" with political events then taking place, it is proper before proceeding further to give the explanation given by Mr. G. Henry Kinahan, of the Geological Survey of Ireland, of such expressions as we find in these annals, which explanation is quite satisfactory. He says, " During the auroras in 1871, I saw here, (Ireland) the lakes and rivers looked as though full of blood." That is at the time of the many fiery red auroras of that year. He suggests that the people of olden times probably had butter and milk for supper, and eating without light, as they did, the color of the red auroras was reflected in them, and hence such expressions as, the butter, the milk, the cakes, the lochs, rivers, etc., were turned into blood.

We have followed the annals down to 685, where we find it recorded that it rained blood in Britain this year, and butter was turned into blood. Here a most extraordinary period of auroras commenced, and we have continuous records of them from 685 down to 692, for a period of seven years. As this period peculiarily interests us from the fact that it first suggested a Saturnian Cycle, we ask the indulgence of the reader for a little deviation. Our first verification of the Jovial Cycle was by comparing the auroral periods with it. Of the early auroras not one failed in falling upon a Jovial Cycle, excepting that of 670 which we then laid aside for future consideration and investigation. But here we fell upon an unintermittent period of at least seven years of fiery auroras, showing both continuous and intense perturbation during all that time. What was the cause of this perturbation, we asked, and the answer came back, "A Saturnian Cycle!" When, from the best data we could command, we had determined the equinoctial points on Saturn's orbit, and calculated the time of his last passage of one of those points, finding it to be in 1863, and then compared it with the time elapsed between the year 685 and 1863 by dividing the time by the length of the Saturnian year, what was our utter astonishment to find that Saturn passed the identical point in the year 685. Hence the perturbation was accounted for, and I considered the hypothesis that planetary equinoxes are the cause of periodical physical disturbances, no longer a hypothesis, nor a theory even, but a demonstrated and verified truth.

The chronicles record the following phenomena during this period.

"A. D. 686, fiery snow fell all night on Easter Monday."

"A. D. 687, great prodigies seen in the sky; fiery dragons were seen, and it rained blood all night."

"A. D. 688, the moon turned into the color of blood on the festival of St. Martins. The Brut y Tywisogion (the Chronicle of the Princes) corroborates this statement of the Scotch Chronicle, but in different words, thus, 'A. D. 688, it rained blood in the island of Britain and Ireland.'"

"A. D. 689, (Chron. Scot.) at the time of the battle against the son of Prida, bloody rain fell in Lagenia."

"A. D. 690, (Brut y Tywg) the milk and butter turned into blood."

"A. D. 690 or 691, after the battle of the Leinstermen with the Ossorymen, fought in King's County, Ireland, in 690 or 691, wherein Foylcher O'Moyloyer was slain, it is said 'a shower of blood fell, and blood flowed for three days and three nights; milk and butter were turned into the color of blood, and a wolf was heard to speak.'"

There is a discrepancy in the date of this battle. The Annals of Cloonmacnoise say, A. D. 688: The Annals of the Four Masters say it happened in 690; another authority says 691; while the Annals of Tighernach place it in 693.

"A. D. 692, (Brut y Tywg) the moon turned into a bloody color." The minor equinox of Jupiter occurred A. D. 691.66. The auroral period in consequence of the Saturnian equinox extended over the whole of the intervening time between the major and minor equinoxes of Jupiter, and including them both.

"A. D. 714, (Chron. Scot.) it rained a shower of honey upon Othan Bee; a shower of silver upon Othan Mor; and a shower of blood upon the Foss of Laighen." The shower of honey probably means an orange aurora, the shower of silver a white, and the shower of blood a red one.

If these auroras happened in Autumn, the usual season of their greatest prevalence, then they preceded the Jovial equinox about eight months, for it took place 715.41.

"A. D. 744, a red crucifix appeared in the heavens after sunset." *Ibd.* The major equinox of Jupiter took place 745.08. Hence this aurora coincided very closely with the equinox.

"A. D. 793, this year dire forewarnings came over the land of the Northhumbrians, and terribly terrified the people; these were excessive whirlwinds; lightning and fiery dragons were seen to fly in the air."—*Anglo Saxon Chronicle.* The fiery dragons here may have meant the tornado spout which often glows with electric light, and appears as if on fire, and may not as usual, mean the shooting corruscations in auroras.

Here we have electric explosions, tornadoes, etc., the invariable concomitants of physical perturbations. The major equinox of Jupiter occurred 792.5. These phenomena therefore succeed the equinox from eight to ten months.

"A. D. 811, (Scotch Chronicle) This was a year of prodigies. A column of light was seen in the heavens; cakes were converted into blood; and blood flowed when they were cut." The minor equinox occurred in 810.22, or about nine and a half months before the commencement of the year 811.

"A. D. 829, In the Chronological History of the Air and of the Weather, published in London, 1749, occurs this statement: 'An earthquake happened at Aix a few days before Easter, and a violent hurricane for several days together; very many trembling fiery-like stars ran up and down the air; great tempests of wind followed." This was just one year after the major Jovial equinox.

I find in a rough calculation for the equinox of Venus—allow for style and precession of the equinoxes—that it occurred about the 3d or 4th of April, A. D. 929. The tempests of wind therefore must have happened within a few days of the Venusian equinox.

"A. D. 850, a column of light shot up to heaven and remained visible to the inhabitants of that place (Repton) for 30 days."—*Florence of Worcester*. Assuming this to have occurred in Autumn, when fully two-thirds of auroras happen, then it came within ten or twelve months of major equinox, which took place in 851.8.

"A. D. 890, (Chron. Scot.) the heavens appeared to be on fire at the Kalends of January." The minor equinox of Jupiter occurred A. D. 891.45; consequently the aurora preceded it by between sixteen and seventeen months.

"A. D. 944, two fiery columns were seen a week before All Hallowtide, which illuminated the whole world."—*Ibidem*. Jupiter's major equinox occurred the next year, 945.85; therefore the aurora preceded the equinox just a year.

"A. D. 979, that same year was seen a bloody cloud often in the likeness of fire. It mostly appeared at midnight, and so in various beams was colored. When it began to dawn, it glided away."—*Anglo Saxon Chron*. This was an extraordinary aurora, or rather auroras, but since they occurred about three years and more before the major equinox, and about two and two-thirds years after the minor equinox of Jupiter; therefore they are not referable to Jupiter. Calculation however shows that a Saturnian equinox occurred early in the year 979, and therefore these auroras exactly fit a Saturnian period.

Florence of Worcester's Chronicle records of the previous year, 978, the following auroral phenomena which appear almost identical with those described by the Anglo Saxon Chronicle as happening in the year 979, namely: "A. D. 978, at midnight, the 18th Kalends of May (April 14th) there was seen throughout England a cloud, sometimes of a blood color, and sometimes fiery. It afterwards broke up into rays of different colors, and disappeared at daybreak." Gaimer, in his History of the English, says, "At midnight as he [the murdered King Edward] lay in the moat, a heavenly light spread itself there. The light was bright—no wonder—; it very much resembled the Sun. This ray came over his holy body; the top of it was in heaven." As the date corresponds with that of the death of King Edward, it is evident that the year 978 was an intense auroral year. But the minor equinox of Jupiter had occurred in 976.35; hence it was too remote—over two years—to have been the cause of the intense auroral display in 978. But the occurrence of a Saturnian equinox in 979 fully accounts for these auroras outside of the Jovial periods, and also for their great intensities.

We have now compared with the equinoctial periods in the Jovial Cycle, the dates of the occurrences of all auroras we have found mentioned in Ancient History and Chronicles prior to the year 1000 of our era. We acknowledge that the language employed by the earlier annalists in the period, is so obscure as to leave reasonable room for doubting whether all the records refer to auroras or to some other phenomena. But those of the latter part of the period are so distinctly and unequivocally stated as to render their identity incontestable. Possibly a few may be apocryphal, but the genuineness of by far the greater portion, is unquestionable. As with the earthquakes, it is surprising to see each aurora, when examined, fall into line and find its place invariably within the perturbed or equinoctial periods of the Jovial Cycle. No, not invariably; for there are a few, a very few exceptions. And what do they do, disprove the theory? No, they confirm it by enlarging it. They prove that the theory was only wrong in assuming there was but one cycle, an exclusive one; and that the principle supposed to be special to Jupiter, is general and belongs to all the planets. The facts of more than a thousand years had been passed under review before one of the exceptional facts presented itself. It was told to stand aside for the present, with the promise that at the proper time it should have a hearing, when we had leisure to listen to the tale it had to unfold. The facts of more than another century are passed under review before other phenomena of similar character presented themselves. These phenomena by the infrequency of their recurrence suggested a longer cycle than the one under investigation. Naturally the longer period was sought for in the next larger and outer planet, one of whose revolutions is equal to two and a half that of Jupiter. The suggestion was, that these intense and exceptional phenomena must be owing to Saturn, and that he at his equinoxes must exert a disturbing influence similar to that of Jupiter. The question, Is this so? was propounded to Nature; and the immediate response was, It is so. The exceptional facts that had no place in the Jovial Cycle, now all fell into line, and found their proper places at the equinoctial points of Saturn. I felt, when the revelation broke upon me, as though my task had been triumphantly achieved, and that henceforth Man held in his hands the keys that unlock the mysteries of one of the great departments of Nature.

Since it is not our purpose to give a history of auroras*, but to verify a Meteorological cycle; hence we will not, as heretofore, when the phenomena were so few, consider every one, and assign to it a proper place in its cycle. We have to pass over time as rapidly as possible; because, since the year 1000 the records are too crowded to do otherwise. However, where a period of peculiar interest occurs we will give it the special attention it merits.

"A. D. 1052, a tower of fire was seen on the night of the festival of St. George, during the space of five hours." The festival of St. George is on the 23d of April. The aurora occurring A. D. 1052.32, and the major equinox A. D. 1053.28. Hence the events are not quite a year apart.

Numerous auroras are recorded in the years 1560 to 1564, the major Jovial Cycle was completed 1563.40.

Brilliant and frequent auroras are recorded from 1568 to 1575. The Saturnian Cycle was completed in 1569 and the major Jovial in 1575.16.

Many brilliant auroras are recorded as having been seen in the years 1621, 1622, and 1623. The major Jovial equinox occurred A. D. 1622.46.

Several very brilliant auroras are recorded as having been seen in the autumn of 1633. The major equinox occurred in 1634.32, or less than six months after. In 1705, 1706 and 1707, a number of auroras are recorded; the most remarkable is that which was seen in Ireland, November, 1707. A Jovial equinox took place near the beginning of 1706.

We now meet with the most remarkable auroral period in history, extending from 1715 to 1720. The major equinox of Jupiter did not occur until 1717.58. The intensity of the period,

---

*Note —Those who wish to pursue the investigation thoroughly are referred to Prof. Frohe's "Nova et antiqua Luminis atque Aurorae Borealis Spectacula," published in 1739; and to M. de Mairan's "Traite Physique et Historique de l' Aurora Boreale," published in 1754.

its early commencement and prolongation to over five years suggests the presence of an additional cause, which is found to be a Saturnian equinox occurring on January 1st, 1716; it consequently both precipitated and prolonged the disturbance.

The most memorable and besides historical aurora of this period occurred on February 23, 1716; the day of the execution of Lord Derwentwater, a Jacobite, that is, an adherent of James Stuart, sometimes called James III, who, by the aid of France and the Tories, made an attempt to regain the throne of England in 1715. From the occurrence of this aurora on the day of his execution, auroras, for a long time, were known in the North of England as "Lord Derwentwater's Lights." It is to this aurora that the almanacs of the last century refer as the "Great and Amazing Light of the North, and which continues to be seen at times ever since." On the 6th of March, 1716, there was another aurora, described in the *London Flying Post*, of March 8th, with an explanation of it which is about as amusing and curious as it well could be made. After his explanation, the writer proceeds to do what was an easier task, namely to show that the aurora had nothing to do with politics. Speaking of the Jacobites, that is, the adherents of James (which in Latin is Jacobus), he says. "Some ignorant people,—whose ideas on such occasions are stronger than their senses,—fancied they saw armies engaged, giants with flaming swords, fiery comets, dragons and like dreadful figures; and others fancied they heard the report of firearms and smelt powder." * * * "The disaffected party have worked this up to a prodigy, and interpreted it to favor their cause, etc."

The last really brilliant aurora of this period occurred February 21, 1718; the maximum, however, occurred in 1716. It is pertinent to remark here that at one time during this intense perturbation an immense sunspot was visible to the naked eye.

In the report of the Scottish Meteorological Society, it is shown that immense rainfalls took place in 1717 and 1718.

The extraordinary brilliancy of the auroras of these periods attracted general attention, and of course, as is always the case, however ignorant and rude the people are, when extraordinary phenomena take place, hypothesis are stated and speculations are rife as to the cause of them. The only hypothesis that has attracted attention is that of Dr. Halley, whose name is perpetuated in the comet whose periodicity he determined. He wrote and published several articles in the Philosophical Transactions of 1715 to 1717, to prove that auroras are magnetic phenomena. Kaemtz, the German Meteorologist, about fifty years ago, hit the true theory respecting auroras, that is, that they are both magnetic and electric. He would have succeeded in explaining them had he known the dependent relation between Magnetism and Electricity. As we have traversed this whole subject in the first part of this work, a repetition of it here would be improper. The auroral observations so far given are those of Europe, where the phenomenon is comparatively rare as compared with North America. The only American observations made during the early part of the Eighteenth Century known to us, are those of Prof. Winthrop, of Massachusetts, incomplete, but covering about 20 years from 1739 to 1758.

The major equinox of Jupiter occurred 1741.30. His record runs as follows: September 12, 1739, extraordinary aurora. January 10, 1741, many sunspots visible to the naked eye. In January and February, many auroras are recorded. March 5, 1741, an extraordinary aurora. September 27, 1741, an intense bright aurora. Minor equinox of 1747.23, or late in March. Only one entry on his record March 1, 1747, "An extraordinary Aurora Borealis."

For the major equinox of 1753.16 the record is missing. For the next a joint Saturnian and a Jovial period, his record stops before the occurrence of the Saturnian in December, 1759. The Jovial having occurred about February 1st of that year. His record is "August 13, 1757, an aurora. None for a great while. Nov. 12, 1757, a remarkable aurora. Sept. 7, 13, and 14, 1759, auroras." Mayer observed a large sunspot, March 15, 1758, whose diameter was equal to one-twentieth part of the Sun. He says: "Ingens macula in sole conspiciebatur, cujus diameter=1-20 diam. solis."

We have to pass over all the intervening auroras till the extraordinary one on Oct. 23, 1804, which attracted attention in both the Old and New World, and in both the Northern and Southern Hemisphere, by its extraordinary brightness, which is said in its luminous effects to have been equal to the full moon. This was a pure Saturnian aurora, since Jupiter's equinoxes occurred as follows: major, 1800.58; minor, 1806.51; while that of Saturn occurred 1804.22.

Assuming that we are correct in assigning Sept. 25, 1871, or 1871.74 for the occurrence of the Jovial major equinox, then the equinoxes that have occurred, and will occur in the present Century, are as follows:

| MAJOR. | MINOR. |
|---|---|
| 1800.58 | 1806.51 |
| 1812.44 | 1818.37 |
| 1824.30 | 1830.23 |
| 1836.16 | 1842.09 |
| 1848.02 | 1853.95 |
| 1859.88 | 1865.81 |
| 1871 74 | 1877.67 |
| 1883.60 | 1888.53 |
| 1895 46 | 1900.39 |

The auroras of the present century have been too numerous to be particularly mentioned. The long period, 1811 to 1820, of perturbation was especially characterized by the frequency and brilliancy of its auroras. The French meteorologist, Biot, sketched several of these, and an engraving of one of them, that of August 17th, 1817, is still found in books to illustrate auroral phenomena.

The major equinox of 1824.30, was notable for the brilliancy, I might say appalling display of fiery auroras. Though but a lad, I remember them well, and the terror they inspired—for I shared in it—amongst the country people in a mountainous region of interior Pennsylvania. Illiterate ministers laid aside the gospel, and made auroras their text, averring them to be the signs and wonders in the heavens that foreshadowed the approaching end. This added to the terror, and drove some people into madness. These auroras continued unusually long after the equinox, for even in 1827 some notable ones occurred. There was then a comparative repose till the Autumn of 1829, six or seven months before the minor equinox of 1830.23, when they again appeared with great brilliancy. Capt. Ross, then on a voyage of discovery in the Arctic Ocean, speaks enthusiastically of the unsurpassable splendor of the auroras in the polar regions in the Autumn and Winter of that year. A long and brilliant auroral era had now commenced, which continued down to 1838. The occurrence of the minor Jovial equinox late in March, 1830, and a Saturnian one in 1833, were the causes not only of the frequency and brilliancy of the auroras, but of the prolongation of their term through the major equinox of Jupiter, occurring in 1836. A precisely similar conjunction will exist in 1877, and therefore unusual bright auroras will occur, probably commencing in the Autumn of the present year, till after the major Jovial equinox in 1883.

On the 7th January, 1831, a remarkable aurora occurred that was seen both in Europe and America.* Humboldt says it was so bright that common print could be read by it. An aurora on the night of the meteoric shower, Nov. 12 and 13, 1833, started the hypothesis that auroras were caused by meteoric Matter. The brilliant auroras of November 17, 1835, and of April 23d, 1836, are illustrated by plates in Bradford's "*Wonders of the Heavens.*"

We can only advert to the brilliant auroras of the major equinoxes of Jupiter, occurring in 1848, 1859, and 1871. Those of 1847 and 1848 were intensely bright. I have been informed by lumbermen in the pineries that they often filed their saws by auroral light in those years. The intensely bright aurora on the night of the solar outburst, September 2d, 1859, was seen all over the Globe, and was accompanied by violent magnetic disturbances, observed in both the Northern and the Southern Hemisphere. Those of 1871 have already been partially given in the extracts from our phenomenal record.

During each of these auroral cycles, synchronous with the major equinox of Jupiter, both polar hemispheres of the sky were almost constantly luminous with auroral light. It was during these periods that it was discovered that auroras are synchronous in both hemispheres; and that when an aurora borealis illuminated the Northern Hemisphere, it was found, invariably, that an aurora australis illuminated the Southern. The skies of both hemispheres were filled with shivering billows and glowing streams of light, shining with a white, and sometimes

---

*Note.—A drawing of this aurora was made by Becker. It has been engraved, and illustrates Mueller's "Atlas Zum Lehrbuch der Kosmischen Physik."

with an orange lustre; but oftentimes glowing with a fiery red, varying in all the prismatic colors of the rainbow

In Canada and the Northern portion of the United States, scarcely a week passes that an aurora may not be observed; faint indeed, but distinct enough to be recognized. They invariably precede and often accompany a storm centre across the Continent. In Europe, however, they are rare phenomena, and are never seen excepting during a general physical disturbance. This is the reason why so few are upon record, which however, for scientific purposes, is all the better; because they hence mark periods of extraordinary physical perturbation, which they do not in America unless only the brilliant ones are taken into consideration. So rare is the sight of an aurora in Europe, that Prof. Nichol, of Scotland, gives it as a curious fact, that most of the writers upon auroras, including Dalton and Mairan, never saw one.

In the comparison of all auroras recorded within the historical period, with the times of the occurrence of the Jovial and Saturnian equinoxes, we find they invariably coincide with one or the other. Hence auroras as well as earthquakes confirm and verify the theory that the cycles of physical perturbations coincide with planetary equinoxes; and hence that the two stand in the relation of cause and effect. Since the test in verifying the theory has been to let it stand or fall by the agreement or non-agreement of the dates when the phenomena took place, with the dates of the occurrence of the equinoxes of two planets; and since it has been ascertained that there is a uniform correspondence in dates between them; hence we are obliged unavoidably to accept as a demonstrated truth, that there are two great Meteorological Cycles, namely, a Jovial and a Saturnian one; and that in each of these cycles there are two points which the planet of the cycle cannot pass without producing physical disturbances accompanied by identical phenomena.

Sunspots taken generally are not visible to the naked eye, and hence but few instances of their being observed are on record. The surface of the sun, however, is seldom if ever entirely free from spots: resembling in this respect the polar hemispheres of the Heavens, which are rarely entirely free from auroras. When however, our Earth and its atmosphere indicate a general and violent physical disturbance, then immense spots are also seen upon the face of the Sun. Not only the size, but the number of the spots are then observed to be increased. These we are now warranted to attribute to either Jovial or Saturnian influence. But whatever obtains at a Jovial or at a Saturnian equinox, must also obtain at the occurrence of any equinox of any of the inferior planets, for our theory is not of special, but of general application. The theory then exacts that at the occurrence of an equinox of any, and of every one of the planets, there must be an increase of sunspots. That it is so, at the equinoxes of the Earth is beyond a doubt; and we feel warranted to say the same of those of Venus; since the average cycles of increase determined by observers of 56 days—half the time between two Venusian equinoxes—indicates that at every second cycle, or 112 days, always such increase takes place. But since 112 days—also a cycle determined by observers—measure the time from one Venusian equinox to another, therefore we feel warranted in saying that a Venusian equinox effects an increase of sunspots.

Observations no doubt will determine the correctness of this deduction, but we have no access to any observations made, and no means of making them ourselves. If this proves to be true, then it will be a singular circumstance, if the elements of Vulcan—which have so far eluded all the skill and ingenuity of astronomers—should happen to be discovered and determined, by his effect upon the Sun.

The earliest mention of sunspots in the chronicles is indefinite, and reads as follows: "It is said that *about* the year 535 the light of the Sun was dimmed for the space of fourteen months." If accidental, it is certainly singular that the chronicler hit so near on a period of great disturbance, when such an event as he describes was possible. His uncertainty as to the exact time qualified by the term "about," indicates that he fixed the period as near as he could, and that he referred to what actually took place. Now, the major Jovial equinox occurred in the year 536.75, or a year and three-quarters after the date given, and a Saturnian in the year 537.45, or not quite two years after.

The next record runs as follows: "In the year 626 half of the disc of the Sun was obscured during the whole Summer." A Saturnian equinox occurred in January, 627, and the major

Jovial in February, 625. The time of the sunspots was intermediate, or rather a year and four months after one, and six months before the other.

The next mentioned are the observations of John Fabricius, of Wittemberg, made in 1610, and published in 1611. Harriot began making observations on 8th December, 1610, when large spots were visible. Galileo observed the spots in January, 1611, and Prof. Scheiner, of Ingolstadt, about the same time. Jupiter's major equinox occurred in September, 1610.

In 1769, within a year of the minor equinox, the celebrated Alexander Wilson observed the numerous sunspots that covered the solar disc in that year, and published his theory respecting their cause, which was accepted for a long while, but is now generally discarded. The elder Herschel now undertook their observation. In 1794 (minor equinox year) he observed many; and a very remarkable one on January 4th, 1801, and only six months after the major equinox.

It is unnecessary to enter into details of special sunspots. According to our theory the minima of sunspots would fall, if Jupiter were the sole cause of them, exactly midway between a major and a minor equinox, that is 2.96 years before and after each equinox. But from the facts presented, it must be evident that Jupiter is not the sole cause of physical disturbances; yet the same facts make it evident that he is the controlling cause; for it is evident from the facts presented, that occur what may, Jupiter's agency is never entirely obliterated. Saturn, at long periods, intervenes and influences phenomena both as to time and intensity, but never obliterates the Jovial period. Mars, the Earth, Venus and Mercury, at shorter intervals, bring their power to bear, but it is only to accellerate or retard the maxima or minima of the Jovial Cycle. We will therefore only offer the average maxima as we find them in "Nature" of Jan. 4th, 1872:

| AVERAGE MINIMA. | AVERAGE MAXIMA. |
|---|---|
| Years 1810.5 | Years 1816.8 |
| " 1823.2 | " 1829.5 |
| " 1833.8 | " 1837.2 |
| " 1844.0 | " 1846.6 |
| " 1856.2 | " 1860.2 |
| " 1867.2 | " 1871.6 |

The reader must however bear in mind that this table is

made up by individuals who supposed there was only one perturbation in a cycle, and that this perturbation had a periodicity of ten or eleven years. Hence they give only one maximum and one minimum in the cycle.

De la Rue and Stewart, from actual observation, give the minimum years as 1833.92, 1843.75, 1856.31, and 1867.12. Wolf and Loewy give the same as 1833.8, 1844.0, 1856.3, and 1867.2. De la Rue, Loewy and Stewart give the maxima years as follows: 1836.91, 1847.87 and 1859.69. Wolf gives them as, 1837.2, 1846.6 and 1860.2.

It will be seen by reference to our table, that the major equinox of Jupiter corresponding to these maxima of sunspots, occurred 1836.16, 1848.02, 1859.88, which correspond very closely to the maxima of sunspots according to De la Rue, Loewy, and Stewart; and also within limits to those of Wolf

Of the maxima, as made up prior to constant observation, not very reliable however, two of the maxima occur near the minor equinox, namely, 1818.37, and 1830.23. This was owing to the influence of the Saturnian equinox; and it is also perceived that the nearest minimum to an equinox is that of 1823.2, being only one and one-tenth of a year from the major equinox of 1824.3. These, however, are made up of irregular observations, and at best can only lay claim to being considered as approximations. It is due to say of Schwabe, of Dessau, that he was the earliest of continuous observers of sunspots, commencing as he did in 1826.

This must close the testimony for showing that it is inferable that sunspots have a connection with the equinoxes of Jupiter, since they have their maxima when Jupiter is at his equinoctial points, and their minima when he is at or near his solstices.

We intended also to have proven that magnetic maxima disturbances have the same relation to the equinoctial, and their minima to the solsticial points of Jupiter, as other phenomena. But as we have, with few exceptions, only averages, the observations, if adduced, would only be presumptive evidence, and not positive proof, that such corresponding relations exist. In Science we must have positive certainty that a thing is true, and not presumptive probability, however strong, that it may be so.

We have examined all the averages as reduced, of magnetic

observations, for a period of forty years, and find a uniform and close correspondence between their periods and Jupiter's positions on his orbit, that is, their maxima with his equinoxes, and their minima with his solstices. Like sunspot observations, only the greater maxima and the least minima are kept in sight in magnetic observations and in the search for the length of the cycle; and hence when from the intervention of Saturn, the minor disturbance became the greater and was displaced by over a year, the length of the next cycle, as indicated, was only a fraction over nine years, while the preceding one had been over thirteen years. Their failure in finding the true length of the cycle, was from not distinctly recognizing two maxima in the cycle with two corresponding minima, and keeping them separate. Hence when, from adventitious causes, the major was shorn of a portion of its wonted energy, as in 1824, and the minor more than correspondingly increased, and the maximum precipitated by one year, as was the case in 1830, they, in mistaking the latter for the former, committed so serious an error as to make the attainment of their object an impossibility.

These averages, however, as far as averages legitimately can prove any thing, bear strong corroborating testimony to the truth of our theory. The following must suffice as to these averages, since it fairly represents their uniform correspondence with the periods of Jupiter. It is a table prepared by Colonel, now General Sabine, of his reductions of magnetic perturbations from 1843 to 1848, both inclusive:

| Year. | Ratio No. of Perturbations | Ratio of Aggregate Value. |
|---|---|---|
| 1843 | 0.60 | 0.52 |
| 1844 | 0.78 | 0.78 |
| 1845 | 0.72 | 0.65 |
| 1846 | 1.20 | 1.15 |
| 1847 | 1.28 | 1.42 |
| 1848 | 1.43 | 1.52 |

By reference to table of Jupiter's equinoxes, already given, it is seen that Jupiter's major equinox occurred about the first of January, 1848. Deducting from this date half the time between the occurrences of his equinoxes, and we find that Jupiter was at his perihelion about the first of January, 1845. His minimum three years then were composed of the years 1844, 1845, and half of 1843 and 1846, while in the maximum, half of the year

1849—not given—should be included. What the observations, thus corrected, would show, is not known; most probably it would be still more favorable to the theory. However, as it stands it is strong enough.

Col. Sabine remarks upon this table as follows: "So general a change in the march of all the magnetic elements, demands a proportionate cause." After pointing out the singular coincidences between the years of magnetic maxima and minima perturbations with those of solar spots as observed by Schwabe in these years, he adds; "The coincidence of the maximum and minimum of solar spots and those of magnetic perturbation demands a cosmical cause depending upon the Sun." Very true; but what does the perturbation of the Sun depend upon? And why should this perturbation have such a regular periodicity? It is evident that no regular periodicity were possible, unless there was a fixed cause for it in the solar system. The physical agent that effects this perturbation is unquestionably Electricity; and hence the cause must be such as developes an unusual amount of Electricity in the Sun at these periods. We have shown that planetary equinoxes,—when the solar magnetic pole to which the planet is then more exposed than on any other part of its orbit, has the effect of a moving magnet upon the latter—; and the greatest angle that the plane of its rotation then makes with that of the Sun are, amongst others that might be named, sufficient causes for the development of an unusual amount of Electricity in both Sun and planet.

Observation upon all general physical phenomena, Telluric, atmospheric and solar, have unquestionably established that they are coetaneous with each other in their appearance; coextended in their duration, and covariant in their energies; and, moreover, that they are inseparable, for they never appear singly. It is hence evident that they either mutually cause each other, or that they depend upon and are effects of an occult and unknown cause. A moment's reflection is sufficient to convince us that they cannot cause each other, for that implies that an effect can interchange places and functions with its cause, which is absurd. We know, from direct observation, that a solar explosion throwing up to a great height immense jets of what is supposed to be incandescent hydrogen gas; and that even the sudden appearance or sudden disintegration of large sunspots instantly sends an electric thrill through the whole Globe, and sets its polar skies ablaze with the fiery aurora. Instances have been observed and are recorded, where, at the moment, or in a few seconds after the occurrence of solar phenomena like these, electric currents have been observed and felt in all parts of the World and at antipodean points; and at night while the electric disturbance continued, the skies of each polar hemisphere were illuminated with unusually brilliant auroras. Now, although we may say, and say truly in a restricted sense, that the solar perturbation is the cause of the electric perturbation of the Earth in such cases, yet evidently it is only a secondary and not a primary cause. The telluric effect also undoubtedly reacts upon its cause in the Sun, and intensifies it, but in no sense can it be said to cause it. The solar phenomenon is therefore an effect of an unknown cause, and of a cause that is not inside, but outside of the Sun. Though we are hence compelled to look for the location of the disturbing cause outside of the Sun, yet reason restricts us to and within the Solar system.

Now it were an easy task to discover and locate this cause, if it were the sole cause. But the complexity of solar phenomena shows the cooperation of minor at least, if not some major causes. Taking sunspots for instance. At times the whole cycle of nearly twelve years duration is broken up into short periods with well defined intervals between, of either no sunspots, or of comparatively few. Sometimes these short periods coalesce and form longer ones. At other times the well defined minor periods are obliterated and the spots about equally distributed over the whole interval between the two permanently fixed disturbances regularly recurring in the Great Cycle; and again at other times the two disturbances with corresponding minima between, are alone distinctly cognizable in the Cycle. These facts are inexplicable upon any other hypothesis than that there is not a sole cause but several cooperating causes concerned in the evolution of solar phenomena, and hence that the cause is not simple but compound, and oftentimes complex. One cause however is of such overwhelming energy that, though often modified and obscured, it is never obliterated, but can under all circumstances be distinctly traced in all these otherwise varied changes. It hence follows that that cause, if any, is the one that is ascertainable; and that we must look to it for the solution of the mystery of physical disturbances; for though not the sole cause, it yet is the controlling one.

Logically we have shown that the cause of solar perturbation cannot be located in the Sun, but outside of it; and though outside of the Sun, yet inside of the Solar System. This cause however is not simple but complex; and in its component elements we have found one of such great energy that it is the controlling cause. If it be the controlling cause, then it is so by virtue of its greater energy; in other words, it is the greatest of the component causes. In our search for causes we are restricted to and within the limits of the Solar System, hence where can we locate the greatest cause, outside the Sun in the Solar System, except in the largest planet of that system? Led by the inductions of Reason, we arrive again, but by a different route, at the planet of Jupiter as the seat of the greatest possible disturbance in the Solar System outside of the Sun.

One immense step towards success was attained when the great physical disturbing cause was definitely located in Jupiter. But the facts observed made it evident that the perturbation was intermittent; hence the question at once arose, Why is the influence exerted by Jupiter upon the Solar System intermittent? An examination of the observed facts clearly indicated that the great maximum disturbance by which the Cycle was first recognized, is not the only one in it, but that there is another well defined period of minor disturbance which divides the Cycle into two equal parts. Since the length of the Cycle corresponds with that of the Jovial year, hence each perturbation marks half of that year.

Another important advance had therefore been made; and now reasoning from analogous facts furnished by the Earth, we are led inductively to fix upon the Jovial equinoxes as the periodically disturbing causes of the Sun and the Solar System. The next step now is to fix the equinoctial points on Jupiter's orbit; and then to ascertain when he was last at either point. This done, it is easy to calculate when he was at either equinox in all Time that is past, or when he will be at all Time to come. The final step can now be taken, namely, to verify that physical events correspond in the time of their occurrence with the dates of Jupiter's equinoxes. A comparison of the dates of every variety of physical phenomena with those of Jupiter's equinoxes, must at least astonish, if it does not convince the most incredulous of the truth of our theory by their uniform and almost invariable correspondence in dates. We leave the matter here for the calm and deliberate judgment of the reader whether we have made out our case or not. For ourselves, without any hesitation and qualification, we accept it no longer as a theory, but as a demonstrated and verified truth that not only the occurrence of Jupiter's equinoxes but those of the other planets are the causes of disturbances in the Sun, and consequently in the whole solar system.

We had intended also to give a verification of our theory by showing that the maximum of rainfalls and of Cyclones—by which we especially mean the tropical hurricanes—also have their maxima at the equinoxes of Jupiter, and their minima at his solstices, but we must forbear. If the evidence we have adduced is not deemed sufficient by any one to prove the correctness of our theory, then no kind or amount of evidence will do so as far as he is concerned. There is no help for him, and he has "to be delivered over to Satan for the crucifixion of the flesh."

We will however state what has been the general results of our investigation of rainfalls and Cyclones with reference to their prevalence at the Jovial equinoxes. We have examined the Scotch tables published by the Scotch Meteorological Society, extending back to nearly the beginning of the last century, and the tables of the rainfalls of England from 1726 to 1869. In Scotland the rainfalls were immensely in excess in 1717 and 1718. The major Jovial equinox occurred in the year 1717.58. From 1717.58 to 1865.81, twenty-four Jovial equinoxes occurred. At fourteen of these the rainfalls within the disturbed period were from five to forty per cent. in excess of the average for the decade in which they occurred. At five, they were from two to five per cent. in excess of the decadal average. At four they were from an average of one to two per cent. above; and only at one, the minor equinox of the year 1759.09, was it below nearly two per cent.

The averages in the tables are made out for the Calendar year.

If there were a possibility of separating the rainfalls so as to show exactly the amount that fell for three years, say twenty months before and sixteen months after, the table indicates that probably every period would show a large excess.

We have fragmentary observations made in the Southern Hemisphere at several different localities, in South Africa, Maritius, India, Australia and Tasmania. They seem to point to the same conclusion, but as they were used to prove the existence of a maximum rainfall in every eleven years, and a corresponding minimum, the years of the minor maximum are omitted, consequently they are unavailable for our purposes.

In Dr. Englemann's tables of continuous observations since 1839, we find the same correspondence, or rather coincidences of the periods of greatest rainfalls with the Jovial equinoxes. We have separated the observations so as to show the rainfall for one year, that is six months before and six months after the occurrence of each equinox. Dr. Englemann gives the average rainfall for thirty years at St. Louis, 44.48 inches. From these tables of monthly rainfalls, we ascertain for one year—that is for six months before and six months after—at the minor Jovial equinox occurring 1842.18, the rainfall was 46.3 inches: for the major that occurred in 1848.2, it was 66.96 inches; for the minor of 1853.95, it was 46.69 inches; for the major of 1859.88, it was 66.86 inches; and for the minor of 1865.81, it was 46.87 inches; for the major of 1871.72, we have not the tables. A comparison of these amounts show a singular fact, probably accidental, that at the major and minor equinoxes the amounts bear a uniform ratio to each other. The amounts at all the minor equinoxes vary only about half an inch, but are about five per cent. in excess of the average for thirty years; while the amount of rainfall at the major equinoxes also is nearly identical, varying only one-third of an inch, but are about fifty per cent. above the average for thirty years.

We have compared the dates of 104 tropical hurricanes that have happened since the year 1675 to 1875, with the dates of the Jovial equinoxes; we find that seventy-nine occurred within twenty months before, or sixteen months after a Jovial equinox. The date of two, those of the 3d and 19th of September, 1804, correspond with a Saturnian equinox; and twenty-three are not found within the limits we have assigned—three years—for the duration of a Jovial disturbance.

Mr. Chas. Meldrum, of the observatory at Port Louis, Mauritius, in the focus of the hurricanes of the East Indies, has made them a special object of observation and study. He finds by averaging the periods, since their occurrences have been noted, that their average periodicity is eleven and one-ninth year. It is well known that there is a sporadic case once in a while of every kind of physical phenomena. Now it is easily seen that one or two such cases, when taken into consideration, will entirely vitiate averages, which at best are only approximations, and make them wide from the mark. Senor Andreas Poey, of the observatory of Havana, Cuba, in the heart of the hurricanes of the West Indies, has also given much attention to them, and has published the most complete catalogues of them we have. He also finds their periodicity to be about eleven years. The same remark applies to his deductions as to those of Mr. Meldrum. The wonder is, that with their method of deduction they come so near the true period. When, however, hurricanes are taken individually, and their dates compared with those of the Jovial equinoxes, it is ascertained that they as uniformly find their places and fall into line as other phenomena do with these equinoxes.

Jupiter's disturbing influence, on an average, extends fully over three years. The perturbation indubitably consists in his imposing upon the Solar System so intensely high an electric charge, that each member of it becomes a veritable *Gymnotus Electricus* swimming in Space, delivering instantly an electric broadside upon the least provocation. The charge, however, is static, for if it were not so, we would have an incessant turmoil for three years, of earthquakes, hurricanes, tornadoes, and rain, hail, and thunder storms. Intense as the charge is, yet, the whole is in equilibrium, for all members of the Solar System are similarly affected. They are in an irritable condition, and the moment any one of them is affected so as to change its electric condition, the equilibrium of the whole system is disturbed, and each member in its own way proceeds to readjust the equilibrium.

The Earth and its Atmosphere, though intensely charged with Electricity, may yet be perfectly tranquil, until either in its own path of its orbit it encounters an exciting cause, or until it is affected by similar occurrences in Mercury, Venus and Mars, when instantly a spasmodic paroxysm ensues either by earthquakes and volcanic eruptions in the Earth, by hurricanes, immense rainfalls and terrific electric explosions in the Atmosphere, or by all combined, synchronous in time and often coincident in place. Now Mercury will bring on such a paroxysm every 44 days, Venus every 112 days, the Earth twice a year, and Mars once in a year.* Hence during the time of the continuance of this excited condition, a paroxysm may, and probably always does, in a modified form, take place when any one of these causes supervenes. When there is an extraordinary conjunction of these causes—as there was in February and March, 1871, and in August, September and October, of the same year—the paroxysm is almost unintermittent for eighty or ninety days, but ordinarily under the influence of a single exciting cause, the excitement expends itself in fifteen or twenty days. The Jovial disturbance, since it lasts for about three years, hence manifests its presence by great violence at the occurrence of a planetary equinox of one of the inferior planets long before and after the occurrence of the Jovial equinox itself takes place. This is the reason why so many energetic phenomena occur eighteen and sometimes twenty months in advance, and twelve to sixteen months after Jupiter's equinoxes.

We have already stated that the Sun is scarcely ever free from spots, and that the aurora is scarcely ever absent from the polar skies; this is true of earthquakes and cyclones. A few sporadic cases of each will occur at all times, but they generally are of a mild form. But these phenomena when they occur with unusual frequency and with tremendous energy, are always found to be synchronous with the physical disturbance imposed by a Jovial or Saturnian equinox, and that the paroxysm is brought about by the influence of one of the minor planets.

Guided by facts we have discovered the unity of physical phenomena and identified the cause whence they emanate. We now know their cause, the fixed time for it to act, the length of

---

*Note.—There is a strongly marked period of disturbance recurring every 23 days. This we attribute to Vulcan. If this conjecture should prove to be correct, then his year consists of about 46 days.

the period of its activity, and the laws that control it. Hence we know how and why the phenomena occur and approximately the time when they will occur. But there is one thing we do not know. We do not know; except empirically, the relation between the cause and the effect, nor the correlation between the effects themselves nor of their mutual interdependence.

A vast field for discovery is now open and within our reach, but to reap the rich harvest, the whole scientific and unscientific world too must change their method of observation. Such facts as those observed accidentally by Messrs. Carrington and Hodgson, independently and at different stations, on the 1st of September, 1859, point out the direction that observation must take in the Future. They were observing a large sunspot when they were startled by a sudden solar outburst. Two patches of most dazzling brilliancy, far exceeding the brightness of the Sun, burst forth and moved with great rapidity—at the rate of over 7,000 miles a minute—over the dark spot. But the most wonderful circumstance is that, the instant the outburst took place, not only their own magnetic instruments were disturbed, but the self-registering magnetic instruments in all parts of the world show the same perturbation at the moment. Moreover, at the same time, "earth currents" appeared on all telegraph lines. In the United States, some operators on the line between Washington and New York, were knocked down. In England, fire flowed from the pen of Bain's telegraph; and in Norway, telegraph instruments were set on fire. That night an electric storm prevailed over the whole globe; bright auroras illuminated the polar skies of both hemispheres: and the town of Chirvan, in the government of Tiflis Asia Minor, was buried by a mountain being thrown upon it during an earthquake.

Similar observations had been made before, and have been since. The researches of Emile Kluge show many synchronisms between earthquakes, volcanic eruptions and magnetic disturbances. Prof. Lamont, of the observatory near Munich, and Prof. Colla, of Parma, Italy, both observed a violent perturbation of their magnetic apparatus on the morning of April 10th, 1842, so extraordinary that both recorded the hour and minute. Afterwards it was ascertained that at the very moment a terrible earthquake occurred in the Grecian Archipelago. Elsewhere we have given the observations of Matteuci, that electric currents of greater tension and greater variability flow through the Earth during rainstorms, and in clear weather also when it is windy or when there are sudden and extreme oscillations of the barometer, than in clear and tranquil weather. We have also given the observations of Berlandier, on the frontier of Texas, and in the State of Tamaulipas, Mexico, upon the increase of temperature of cold water springs, often two days in advance of the appearance of a tropical hurricane upon that coast. In the present chapter we have given the synchronisms between magnetic disturbances observed at Havana, and the enlargement, sudden change and disintegration of sunspots, while synchronous tropical hurricanes, accompanied with earthquakes, were raging on the adjacent coast of Florida, or amongst the adjacent islands.

Now, since magnetic disturbances,—as we have shown elsewhere,—are but electric disturbances; and since sunspots are synchronous with sudden and violent electric currents in the Earth; and electric currents are synchronous with oscillations in atmospheric pressure, with rain and wind storms, with hurricanes and with earthquakes, therefore we know that the bond of union between all these phenomena is Electricity. The periodicity of the phenomena is owing to the ebbs and flows of Electricity; and the ebbs and flows of Electricity in the Solar System ensues from the peculiar relations that subsist between the Sun and each planet at its equinoctial and solsticial points.

Let this great truth be once indelibly impressed upon the popular mind, and let a definite and true conception of the relations between physical phenomena and their laws and causes be formed, so as to displace the vague notions now generally prevalent, and a stimulant to research and an impulse to progress will be given whose extent and limits no one can now foresee, nor foretell.

# CHAPTER II.

*Venusian, Mercurial, Martial and Vulcanian Cycles.*

The Jovial Cycle furnishes the principle around which as a *nucleus,* all meteorological phenomena crystallize. Hence we have endeavored to lay its foundation deep upon a firm and solid bed-rock of facts. There it must permanently stand, unshaken and immovable, amid all the wrecks of Time, unless the authenticity of the facts upon which it rests is overthrown, so that their *dictum* no longer can be accepted as truth.

But if true, then it is not barren, for truth is always prolific, and it always bears fruit of its own kind. The small kernel first discovered in Jupiter, will, under the operation of the laws of Nature, develope and expand into a gigantic tree whose branches will fill the skies. In Nature there is no isolated principle. None stands there solitary and alone: for Nature does not operate by special, but by general laws. Hence, if Jupiter were the sole planet in the solar system, though he might represent a general principle that obtains universally in other solar systems; but in our system he apparently would represent only a special principle. But placed as he is here, in companionship with co-planets, he is subject to the operation of the same law that governs the system; and in whatever manner that law affects him, it likewise affects his co-planets under similar conditions. If in passing the equinoctial points on his orbit, his own electric condition is intensified, and he reacts upon the Sun and intensifies its electric condition, then the same effects in degree are produced when every other planet passes its equinoctial points.

Since facts unquestionably prove that the Earth and the Sun, and even Jupiter himself, are disturbed when he passes his equinoctial points, hence facts must also exist that prove this to be the case with all other planets under similar circumstances. That this is the case with the Earth at her equinoxes, has been known and admitted so long that " the memory of man runneth not to the contrary;" and that she likewise communicates her infection to her coplanets, cannot be doubted so long as the physical law is in force, that in a group of insulated bodies the electric condition of no one of the members can be either eased or intensified, without decreasing or increasing in proportion the electric tension on all the members of the group. Such are deductions warranted by general principles; and this can be either verified or disproved by the records of facts which Time and the diligence of observers have accumulated. These records we will now proceed to examine, and see whether the Earth and its Atmosphere, and the Sun himself, give any evidence of disturbance when Venus or Mercury pass their equinoctial points.

To impress the condition under which violent paroxysms occur more indelibly upon the mind of the reader, we will recall the fact repeatedly stated in regard to Jupiter's long perturbation of nearly three years. Though the frequency and energy of Telluric and atmospheric paroxysms during this period indicate an abnormal electric condition, yet the great and terrific paroxysms invariably occur when the equinoctial perturbation of one of the inferior planets is superimposed on the Jovial. Take any Jovial period, and the largest sunspots, the most brilliant auroras, the most appalling earthquakes, and the most terrific Cyclones, invariably occur at our own equinoxes, or at those of one of the inferior planets. The same is true of our own equinoctial disturbances, and those of Venus also. The most violent and terrific paroxysms occur when another planetary equinox falls within the given period of perturbation; for instance, at an equinox of the Earth and one of Venus; one of Mercury and, above all, what I consider the equinoctial disturbance of Vulcan. For this short period, as we will show at the proper time, like the long Jovial period, is never obliterated, and invariably at its fixed time manifests its presence, and under proper conditions, with terrific energy. Its constancy and its energy led to its discovery.

The reader therefore perceives that the question of the time for the occurrence of physical paroxysms is not a simple one that can be astronomically determined by merely determining its exciting cause, but is a compound and often complex one, depending upon the approximate coincident occurrence of one or more exciting causes. Probably in but few cases is a phenomenon purely owing to one cause. It is certain that all violent ones are the combined effects of several cooperating causes. For example, in Jupiter's Cycle—lacking only 51 days of 12 years—he fixes two periods at which physical perturbations will occur. The duration of each of these is estimated at three years, though the perturbations of the magnetic needle, and the prevalence of sunspots show that the influence is sometimes very sensibly felt for a longer period. These periods are separated by intervals of comparative tranquility of the same length as themselves, in which physical phenomena are rare, and what occur are generally of a mild character. But the periods of perturbation abound in frequent and violent physical disturbances, often unintermittent for four or five months, and then intermit for as long a period, when they again recur.

It is during these periods, as we have shown, that nearly all the terrific physical phenomena recorded in history have occurred. They are brought about in this way: During the period of Jupiter's excitement—(the Ancients called it Jupiter's wrath)—the Earth obtrudes by her equinoxes six periods of excitement, Venus nine, and sometimes ten, Mercury twenty-four or five, and Vulcan about forty-eight. However, as nearly two-thirds of the equinoxes of Venus fall near to or within twenty-five or thirty days of the Earth's, so in the three perturbed years we have only seven or eight seasons strongly marked by violent paroxysms, though all the elements of perturbation are more or less active during the whole time. During the disturbed period that culminated in the Autumn of 1871, we had only six such seasons so strongly marked by their violence as to give them prominence; namely: first, September, October and November, 1870; second, January, February and March, 1871; third, June and July, in same year; fourth, August, September and October, of same year; fifth, March and April, 1872; and sixth, August, September and October, same year. The first of these periods was purely Jovial and Telluric; the second, Jovial, Venusian and Telluric; the third, Jovial and Venusian; the fourth, Jovial, Martial, Venusian and Telluric; the fifth and sixth, were Jovial, Venusian and Telluric. Each period, of course, was besides complicated by several Vulcanian and one or two Mercurial disturbances.

It is hence perceived that paroxysms are precipitated at any time during a Jovial disturbance when an equinox of one or of several minor planets supervene, and that such paroxysm endures while the adventitious disturbing cause lasts, when comparative tranquility again ensues until another such disturbing cause supervenes. This is the reason why the terrific phenomena so characteristic of these periods are so widely distributed as to time, some coming a year and a half before and others as long after the occurrence of the Jovial equinox.

The same principle obtains in Telluric and Venusian equinoxes. The violent phenomena to which they give rise, hardly ever are synchronous with them. If they are, then it is because an equinox of Mercury or of Vulcan for the time being coincides, or nearly so, with the Telluric or Venusian equinox. Hence phenomena in the Telluric and Venusian cycles are also distributed over the entire period of the perturbation caused by them, because they depend for their occurrence upon the intervention of a secondary cause. For instance, in the Northern Hemisphere, the tropical hurricanes, which plainly are owing to changes that will be completed at the Autumnal equinox, by the intervention of secondary causes, often occur in the latter part of July, and continue to occur to the end of October. In the Southern Hemisphere, owing to the same causes, they begin to appear in the latter part of January and continue till after the middle of April. No exact coincidences of physical paroxysms with Telluric and Venusian equinoxes can hence be expected, though frequently, from the causes we have indicated, such coincidences do occur. The latest case is that of the terrible tornado in Alabama and Georgia, of March 20, 1875, synchronous with the Telluric and Vulcanian equinoxes of that date. But, though phenomena exactly coinciding with Telluric and Venusian equinoxes are rare, yet with Mercurial, and especially with Vulcanian, they almost invariably coincide to the day. It is to be regretted that our information of the latter planet is so meagre, being confined exclusively to its existence, its nodes, and his probable periodicity. We are now prepared to proceed and verify the existence of a Venusian cycle, consisting of 224.7 days, or .615186 of a year, with a physical disturbance at the end of 112 days.

## VENUSIAN STORMS.

Under this head we shall include all the storms falling within the Venusian equinoctial period; though in their production, Jupiter, Mars, Earth, Mercury and Vulcan have participated. Wherever we conveniently can, we will state what planets conjoined in the production of the phenomena.

The records of storms are abundant in history, but no date except the year is given prior to the year 1000. Jupiter extends his influence over three years, hence all Jovial phenomena can be identified when the year is given; but Venus extends her influence hardly over two months; hence, if the month is given, we can approximately identify them, but there is no certainty unless we have the day given.

It is stated that in the year 944, 1,500 houses were destroyed by a tempest in London. No other date is given. In the year 1090, it is recorded that a violent tornado overturned 606 houses in London alone.

The first specific date of a storm we find, is the one that occurred on the 5th of October, 1091, in England. The account states that it came from the South-west, and that the sky was terribly dark. Many churches were destroyed, and that in London over 500 houses fell. A Venusian equinox occured October 12th, 1091, hence corresponding within seven days. This date probably is old style: hence if corrected for style and precession of the equinoxes, it will still correspond within 15 days of the date of the equinox.

It is said by Hollingshead that, in the month of January, 1382, when the queen of Richard II. came from Bohemia, on her setting foot on shore an awful storm arose, and her ship and a number of others were dashed to pieces in the harbor. A Venusian equinox occurred February 5th, 1382; though the day for January is not given, yet as this is old style, 17 days have to be added for style and precession of the equinoxes, to whatever day in January it may have occurred; it hence may coincide with the equinox, or at least must approximate very closely to it. Hollingshead relates as a singular coincidence that Richard's second queen also brought a storm with her to the coast of England, in which the King lost his baggage, and many ships were destroyed; but the only date given is the year. 1396.

Many storms are recorded, but none that we could find with day given, until we came to the wreck of the *Mary Rose*, 60 guns, going from Portsmouth to Spithead, and all on board, perished in a storm July 20th, 1545. This however was not caused by a Venusian equinox, it having occurred 52 days after the preceding, and 60 days before the succeeding Venusian equinox. Correcting the date for style and precession of the equinoxes and it corresponds to August 4th, 1545, of the epoch of 1875. A Mercurial equinox occurred on August 4th, 1545. If our determination be correct of the length of the Vulcanian Cycle, then that cycle completed itself on the 31st of July, 1545, or five days before.

The general storm that prevailed in England, and in fact throughout Europe, the day that Cromwell died, September 3d, 1658, corresponds with the following equinoxes: Venus, August 19th; Mercury, September 1st; Vulcan, September 2d; and Earth, September 21st. If the date is corrected for style and precession, it brings it within five days of the Autumnal equinox. I consider it a Telluric phenomena, intensified by the approximate conjunction of so many planetary equinoxes.

The next specific date is September 1st, 1691, a storm in which the *Coronation*, 90 guns, foundered off the Ramhead, and the *Harwich*, 70 guns, wrecked and all on board perished. No Venusian equinox comes within 40 days of this disaster. A Mercurial equinox occurred August 31st, 1691, or one day of date given; but correcting the date of the disaster for style and precession, and it corresponds to September 14th, 1691. The Vulcanian cycle completed itself exactly on this day

We now come to what is called the "Great Storm," the one that occurred on the 26th and 27th of November, 1703, in England and throughout Europe. The loss sustained in London alone was estimated at £2,000,000 sterling. The number of persons drowned in the floods of the Severn and the Thames, and lost on the coast of Holland, and in ships blown from their moorings, and never heard of afterwards, is estimated at 8,000. Twelve men-of-war, among which were the *Sterling Castle*, *Mary*, *Northumberland*, of 70 guns each, were lost on the Goodwin; the *Vanguard*, 70 guns, sunk at Chatham; the *York*, 70 guns, near Harwick, all on board perished except four men; on the *Resolution*, 60 guns, wrecked on the coast of Sussex ; and on the *New Castle*, 60 guns, at Spithead, 193 were drowned ; on the *Reserve*, 60 guns, at Yarmouth, 173 persons perished ; and on the smaller vessels, altogether 1800 were lost in sight of shore.

In Kent alone 17,000 trees were torn up by the roots. Eddystone lighthouse was destroyed, and in it the ingenious Winstanley, the contriver of it, and all the persons with him perished. The bishop of Bath and of Wells and his lady were killed in bed, in their palace in Somersetshire. Immense numbers of cattle were killed and drowned. The devastation on land and the loss in shipping was immense. The date of this storm, corrected for style, etc., is December 8th and 9th, 1703. The equinox of Venus occurred on November 9th, 1703 ; or 29 days from the date as corrected. It probably therefore had little influence. A Saturnian disturbance was however prevailing and in full energy at the time, and an approaching Jovial equinox was only 18 months distant. An equinox of Mercury occurred on the 9th of December, or on the second day of the storm, and a Vulcanian one* on the 8th of December, the day that the storm commenced.

An awful storm prevailed in the North of England, Oct. 29, 1775 ; many ships destroyed, and four Dublin packets wrecked. Equinox of Venus Oct. 27th, 1775 ; of Mercury, Nov. 1st, and of Vulcan, Nov. 6th

Hurricane at Surat, East Indies, April 22, 1782. A Jovial equinox occurred about five months after this cyclone. Venusian April 13th ; Vulcanian, April 29th, and Mercurial May 4th.

October 3d to 18th, 1780, a terrible hurricane, or perhaps several, prevailed in the West Indies. On the 11th of October it

---

*NOTE.—The reader's attention is particularly called to what we have provisionally named the Vulcanian Cycle, recurring every 23 days; and to the surprising exactitude with which it almost invariably hits the dates of every violent physical phenomenon. We had with much labor calculated approximately its exact period, but afterwards discovered, if we discarded the fractional parts of both our own year and that of Vulcan,—considering the former as 365 days and the latter as 46 days, the number 23, that is, the interval between his equinoxes, measured the phenomenal periods for over 300 years, without any appreciable deviation. We will not positively assert that our determination is correct, but that it is an exact, well and sharply defined Meteorological Cycle, admits of no doubt. We appeal to facts, and to them alone, to decide this matter.

raged most terribly at Barbadoes; over 4,000 of the inhabitants lost their lives. A twelve pounder was taken up and carried 140 yards. The following vessels of the British navy were lost in this hurricane: *Thunderer*, 74 guns; *Sterling Castle*, 64; *Defiance*, 64; *Phœnix*, 44; *LaBlanche*, 32; *Laurel*, *Shark* and *Andromeda*, 28 each; *Deal Castle*, *Penelope* and *Scarborough*, 24 each; *Barbadoes*, *Camelcon*, *Endeavor*, 14 each; and *Victor*, 10 guns. This hurricane was at Havana on the 16th, and at the Bermudas on the 18th. The following approximate conjunction of planetary equinoxes were taking place. Earth, September 21st; Venus, September 29th; Vulcan, October 1st; and Mercury, October 9th. It ought to be noted that tropical hurricanes originate at sea, and move very slowly, so the exact date of their origin is hardly ever known; the date given for them is that when first observed. The same hurricane has often been seen for twenty consecutive days.

We may as well mention another terrible hurricane by which Barbadoes was laid waste on the 10th of August, 1831. Over 2,500 of the inhabitants were killed, and over 5,000 wounded. A piece of lead, weighing 4,000 pounds, was lifted and carried over 300 yards. The Jovial perturbation of the year previous, reinforced by a Saturnian, still prevailed. Venus had passed her equinox on the 29th of June; but the Mercurial equinox occurred August 4th, and a Vulcanian August 8th. This hurricane, it is perceived, was one of unusual violence and of long duration. On August 10th, the day that it devastated the Barbadoes, a severe earthquake occurred in the East Indies. Col. Reid, Governor of the Bermudas, records in his journal: "That during the 11th and 12th of August, the Sun was of a bluish color, and its light unusually dim;"* a fact that will be appreciated by physicists three or four generations hence, as showing what is now shown by the spectrum, that Light passes by way of *blue* into Electricity, by way of *red* into Heat. But this is forbidden ground for us at present.

On the 8th of November, 1800, a general storm prevailed throughout England, doing immense damage both on sea and

---

*Note.—I witnessed the same phenomena on the same days, in the State of Maryland, where I was engaged in teaching. A similar phenomenon occurred a few weeks later, only the Sun was purple instead of blue.

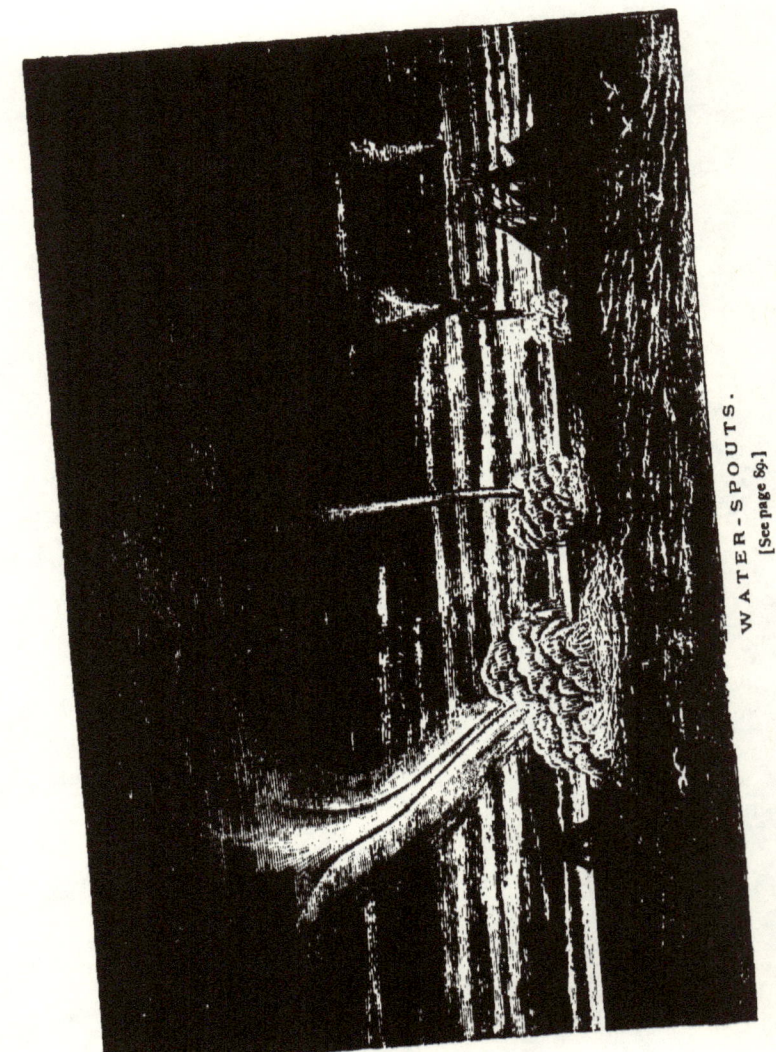

WATER-SPOUTS.
[See page 89.]

land. Jupiter's equinox occurred in June of this year; that of Venus, October 24th; of Vulcan, November 6th; and that of Mercury, November 17th. In Europe storms generally come from the ocean; hence generally appear later than their exciting causes.

December 16th and 17th, 1814, "a tremendous storm throughout Great Britain and Ireland, by which immense damage was done and many ships were wrecked." Venusian equinox, November 21st; Vulcanian, December 18th; Mercurial, December 23d.

December 12th, 1822, "a storm in Ireland, particularly in the vicinity of Dublin, many houses thrown down and vast numbers unroofed. A Venusian equinox occurred November 23d, 1822; a Vulcanian, December 3d, and a Mercurial December 2d, 1822.

The genus Cyclones embraces all kinds of rotary storms. Sand-spouts of sandy deserts; cloud and dust spouts on land; and water-spouts at sea, are species of Cyclones. When a tornado with its dust and cloud-spout passes from the land over a body of water, it becomes converted into a water-spout, and *vice versa*. But there are distinct varieties of each, in which the conversion does not take place; *e. g.*, the water-spout of a calm atmosphere and tranquil sea; and the dry whirlwind in calm weather carrying up an immense column of dust.

On the 27th of June, 1827, occurred, on the coast of Sicily, in the Mediterranean Sea, those grand, yet as may well be conceived, awe-inspiring water-spouts represented in our engraving. The following conjunction of equinoxes occurred at this time; namely, Vulcan, July 4th; Venus and Mercury both on June 30th.

January 12th and 13th, 1828, "an awful storm in England, and along the British coast. Many vessels lost, and thirteen driven ashore at Plymouth alone." There was a Venusian equinox on the 13th of January.

*The Black Sea Storm.*—From the 13th to the 16th of November, 1854, a terrific storm raged in the Black Sea. Eleven transports (English) were totally wrecked and six disabled. The new steamer *Prince* was lost with 144 lives, and a cargo worth £500,000, indispensable to the army in the Crimea. The loss of life in the other vessels is estimated at 340. Except Jupiter who had passed his equinox about eleven months previous to the date of this storm, but two other planets contributed to this phenomenon, namely, Vulcan and Venus, the former passed his equinox on the 10th, the latter on the 15th, or the day before the storm ended.

According to Professor Dewey, a violent storm raged on the American Lakes on the same day, namely, November 13th, 1854.

I have so far followed in detail an English list. Prof. Müller, in his *Lehrbuch der Kosmischen Physik*, mentions a hurricane that commenced on the 23d of December, 1821, and continued until the 27th. For four days it swept over Europe like a single whirlwind. When it struck the peaks of the Spanish mountains and the Maritime Alps, it broke up into smaller storms, which rushed with devastating fury through all the valleys, smaller in diameter but of equal energy to the original hurricane. This hurricane was similar in character to that of 29th November, 1836, and originated in the Atlantic. I have not been able to find a trace of it in America, yet like the latter storm, it may have originated here and traversed the broad Atlantic. Under these circumstances, it is not to be expected that its date will correspond very closely with its exciting causes. It is very remarkable that seven days before its appearance in Europe, namely, on the 16th of December, 1821, an extraordinary conjunction of the following planetary equinoxes took place, namely, Venus, Mercury and Vulcan.

The following is the storm referred to above: Johnston, in his *Physical Atlas*, says, "One of the most dreadful hurricanes on record in the Temperate Zone, whether on account of injury it caused or the extraordinary extent to which it raged, was that of November 29th, 1836. According to the best information it began in America, crossed the Atlantic, and reached the European coast on the parallel of the English Channel. The first accurate account shows it on the 23d of November, on the east coast of North America, in the parallel of the St. Lawrence. It continued its way over the Atlantic, and approached Land's End on the 29th of November, at 7.45 A. M.; it was at Plymouth at 8.30; at Exeter 9.30; Weymouth 10.15; Poole 10.30; Farnham at 12, noon; London 12.30 P. M.; Island of Walcheron at 1; in Düsseldorf at 2; in Hamburg and Berlin at 6; and in

Koenigsberg at 9 P. M. The storm raged most on the coast of France and Belgium. At Ostend, scarcely a house was left unroofed, and the demand for tile in consequence became so great that the price for 1000 arose from 16 to 30 florins."

The following is ascertained to be the dates of the planetary equinoxes of that date: Mercury, November 22d, and Vulcan, November 23d, 1836, the latter the exact day that the storm first attracted notice on the Northeast coast of America.

We intended to have taken up the British list of wrecks and verified the cycles by them, but it would be too tedious: besides, if the evidence adduced, and that to be adduced, do not prove the existence of these cycles, what evidence will? We have already given a list of the wrecks wrought by the special storms we have considered, we will however give one more, because it will be recalled by many readers. It occurred in the Bay of Biscay, on January 11th, 1866, and produced a deep sensation both in Europe and America at the time. We allude to the foundering of the steamer *London*, on her way to Melbourne, carrying down 220 persons; among them Prof. Wooley, principal of the university of Sydney; and the tragedian, G. V. Brooke. On the same day the steamer *Amalia* also went down with a cargo valued at £200,000. An unusual atmospheric perturbation pervaded the entire globe at the time. In the latter part of December, and early part of January, violent storms are mentioned in the Chinese Sea; and on the 26th and 28th of December, 1865, violent gales on the American Lakes and Atlantic coast, are recorded. In a summary of the English list of wrecks, we find this remark: "Many wrecks and great loss of life during the gales from the 6th to the 11th of January, 1866." There must have been an adequate cause for this extreme and widespread atmospheric commotion, and for its violence and protraction. The following are ascertained to be the exciting causes: Jupiter passed his minor equinox about November 1st, 1865; a Venusian equinox occurred December 10th; a Vulcanian December 31st, 1865; and a Mercurial January 9th, 1866.

In passing from European to American storms, we recall attention to the fact already stated, that European storms, as a rule, originate on the Atlantic Ocean, and hence exist for days and even a week before they manifest themselves on the European coast. We have given an illustration of this fact in the great storm of November, 1836, traced by Mr. Johnston to New Foundland, where it was fully developed on the 23d, or six days before it struck the shores of Europe. The same remark applies to the tropical hurricanes of America. They originate in the tropical sea, many degrees east of the Windward Islands, and exist for days before they strike those islands, and then from five to ten days often elapse before they strike the coast of Florida. The tropical hurricane that struck the coast of Nova Scotia with such terrible fury on the 24th and 25th of August, 1873, affords an illustration of this remark. The *Crest of the Wave* encountered it on the 13th of August, in N. latitude 14° W. longitude 27° as a severe gale. The next definite account of it, is on the 18th, from there the Signal Office located it daily until it struck the coast of Nova Scotia on the 24th, and for days afterwards, for it is supposed that the severe storm that passed over England on the 31st of August, and struck the coast of Norway on September 2d, was this identical storm. Under such circumstances it is seen to be almost impossible to trace this class of storms to their exciting causes. In the case of this storm, however, this can be done accurately.

The Telluric equinox was uquestionably, as well as the Venusian, a factor in the cyclone. The immediate exciting cause was Mercury, who passed his equinox on the 13th, the very day that the *Crest of the Wave* met this storm, then only a severe gale. On the 19th of the month, it was intensified by the occurrence of a Vulcanian equinox, and on the next day, the 20th, by a Venusian. Hence its terrific energy whilst opposite Nova Scotia; for its center did not strike that coast by from 100 to 150 miles.

The loss of life was terrible, and that of property immense, estimated at the time at from $4,000,000 to $5,000,000. The number of houses destroyed was about 900. The Signal Service Office says: "Any endeavor to estimate the value of the property destroyed, is attended with great difficulties, but may be within bounds, if to the $700,000 damage done to wharves and crops, is added $1,000 for each building, $2,000 for each of the larger vessels, and $1,000 for each of the smaller ones, which would give a total of $3,500,000; a sum that it will be seen may easily be far below the truth." The Signal Office Report further states: "That 1032 ships, of which 435 were small fishing schooners, are known to have been destroyed during the 24th and 25th of August, in the neighborhood of the Gulf of St. Lawrence and the Atlantic shores of Nova Scotia, Cape Breton, and New Foundland. On the other hand over 90 vessels were destroyed by this hurricane in its passage over the ocean before it reached Nova Scotia, making a grand total of at least 1123 vessels destroyed within a few days by its power. Two hundred and twenty-three lives are definitely reported to be lost, and the moderate estimate of the numerous cases in which whole crews have been lost, swells this number to nearly 500; and if to this is added the loss of life on land, and the loss in the earlier history of the cyclone, the grand total amounts to at least 600 lives."

This lamentable disaster shows the necessity of elementary knowledge in the first principles of Meteorology more clearly and convincingly than any argument we can adduce, enforcing what it teaches in a manner more powerful than any words would that we can command. Here is a chance for teaching Natural Science in our elementary schools; but let no philanthropist, with more zeal than knowledge, commit the egregious blunder of attempting to teach it before it is discovered; for what now passes for Meteorology is less like that Science than vegetable oysters are like real oysters. *Better teach nothing than teach error.* Elementary books say, "that the wind at the surface of the Earth will tend from the colder to the warmer region, and from the place where there is least vapor to where there is most vapor," when the facts fully as often contradict as sustain this theory. For instance, the warm wind flowing northward, out of an area of a southern high barometer, flows from a warmer to a colder region, and as it advances northward, has its own temperature reduced as the thermometer shows; and the moist winds of the ocean, which every where carry moisture into the interior of continents, watering them by producing rain there, flow as often and as long from the moist region over oceans to dry regions over continents, as the dry air over continents flows into the moist air over oceans. The truth is the principle is false; for as we have shown in Part I, atmospheric movements have nothing whatever to do with either the temperature or hygrometric condition of the Atmosphere. These movements are initiated and controlled by general laws of an entirely different character, as we there demonstrated.

Amid the general zeal for applying Science to facilitate the operations, and for multiplying and cheapening the products of every kind of industry, we have not seen any practical plan for the application of Science to the protection of those products afterwards against the dangers and risks of the elements to which they will be exposed in flood and field, on land and water, either before they are finally garnered, or distributed to the points where needed. Here we see hundreds of lives lost, and from four to five millions of dollars worth of property engulfed and swallowed up in a few hours, and most of it recklessly exposed from sheer ignorance of the first principles of Meteorology. Surely the declaration of the prophet, "the people perish for want of knowledge," is as true in our Age—notwithstanding from self-conceit and pride we maintain the contrary—that it was 2,500 years ago. The sufferers, in this instance, are not to be charged with culpable ignorance, for they were fully as much enlightened on this subject as this Age is; but the fault is with the thought, or rather want of thought, and mental habitudes of our Age laboring under the delusion that Nature is a sealed, not an open book, which every one can read and understand for himself.

Meteorology is that Science which takes cognizance of all atmospheric phenomena and changes, and by investigating their nature and character, ascertains and determines their causes. In this sense, as a Science, it is therefore not yet born, for what now passes for it, is even incapable of explaining the Past, how then can it be expected to divine the Future?

The economy of Nature is perfect, but books, the spectacles, with distorting glasses, through which we look at it, prevent us from perceiving its beauty and symmetry, and from conceiving, admiring and adoring its simplicity, grandeur and sublimity. In it physical causes are so organized and arranged that they produce proper variety in their effects. Heat and cold, rain and sunshine, we must have, but neither continuously. We must have alternations of each, systematically distributed so as to give the proper variety and diversity to the seasons; and this is attained to the utmost perfection by the fixed planetary arrangement Nature has ordained to bring about these changes at regular intervals. The Sun she has ordained to shed abroad the primitive form of Energy, Light: but uniform energy distributed by a body of uniform potency, and in unvarying quantity, would soon produce stagnation in the Sun himself, and both in the planets and in their æriel oceans, unless there were exciting causes that not only at times aroused the planets to greater activity, but influenced the Sun himself constantly to modify, diversify and vary his energy. These exciting causes are the equinoxes by which a planet alternately renews its electric vigor; now by the North, and anon by the South magnetic pole of the Sun, thus alternately imbibing electric energies that, we have good reasons for believing, differ in many essential properties.

We are endeavoring to prove the reality of these exciting causes, and hence are not yet properly justified to make a practical application of the principles they inculcate. But suppose they had been accepted as scientific truths at the time this cyclone occurred, what would have been the deductions drawn from them for the guidance of those on land as well as on water? Mercury passed his equinox on the 13th of August. Now, under ordinary circumstances, that is when the Earth is not laboring under any other equinoctial disturbance, Mercury's equinox, though it would produce a minor disturbance, such as a gentle rainstorm, yet it would not give rise to anything calculated to excite serious alarm. But the Earth was already feeling the effects of her own approaching equinox; and that of Venus had already advanced so far as to culminate in a week. Besides, there was Vulcan hidden in the fiery beams of the Sun, giving, at short intervals, such ponderous strokes as Jupiter might envy, he too would get in his blow the day before Venus. Something quite serious was therefore to be expected, and all possible precautions would have to be taken to avoid danger. Such would have been the deductions; now what are the facts?

The Signal Service, in their *Monthly Weather Review*, say: "The areas of low barometer traced on Map No. 2, were accompanied with slight disturbances, while in the western portion of the United States, but no marked change in the weather occurred until the 11th of the month, when an area of low barometer marked No. 6, was first observed in Kansas.* This storm moved slowly to the eastward with cloudy weather, rain and light to fresh winds, the winds increasing as the centre approached the Atlantic coast, and finally produced the northeasterly gale which occurred on the middle Atlantic and New England coast, on the 14th and 15th."

In the Signal Office Report, 1873, (page 1026) this storm is again referred to as follows: "The depression passing through Delaware on the nights of the 13th and 14th, was apparently accompanied by two, if not by three or four, smaller storm centres, which being of the nature of tornadoes, seem to have done considerable damage, both by wind and rain, in eastern Pennsylvania and Maryland. The most severe winds were reported from the New Jersey coast," etc.

In the official report at Boston, Mass., (*ibidem* p. 241) is this item: "August 13th–15th, 1873—Storm very severe outside. Many vessels staid in port; and a few small boats were washed from their moorings."

Report from Portland, Maine, (*ib.* p. 283): "August 14th and 15th, 1873—Gale from the northeast at this station, but a heavy swell coming in from sea."

In the disasters of shipping on the Great American Lakes, (*Ib.* p. 1044) we find: "August 13th, 1873—Barge J. D. Morton, during the rough weather on Lake Erie, broke loose from tow of steam barge *Sun*, and went to pieces; Scow *Senator* had her foretop mast broken and lost canvass during a squall on Lake Erie." "Schooner *Caroline Marsh* lost 8,000 to 10,000 feet of lumber of deck load by a gust of wind while off Oswego, August 15th. Brig *Pilgrim* lost her topsail yards during storm off Chicago. August 19th—Schooner *J. Rigley* lost part of her canvass off Forty-Mile Point."

---

*Note.—I was present at the birth of this storm on Sunday afternoon, August 9th, in Middle Park, Colorado, at the western base of the Snowy Range. We were under the western rim of the cloud canopy, and encountered a heavy rain from 4 P. M. to 2 A. M. next morning. To the west the sky remained clear, but a fearful thunder-storm raged almost above our heads, on the Range. At Boulder, situated forty miles from the Range, at the eastern base of the mountains, we found the same storm had come down from the mountains and passed eastward on the Plains, on the morning of the 10th.

The last accident occurred on the day of the Vulcanian equinox, and this is the last accident by storm mentioned during the month. It should be remarked here that planetary equinoxes produce high as well as low barometers, as we will show hereafter. A high barometer is a down-pour of air, as we have shown in Part I, from the surface of the ærial ocean, and a low barometer—a storm centre—is an upheaval of air from the Earth. There cannot be an upheaval without a down-pour, and *vice versa*. It so happened that the very high barometers that had successively come from the Northwest, but driven back by the Nova Scotia Cyclone three different times, covered Canada and Ontario for nearly the balance of the month, that is from the 19th to the 28th. This is why no storm centre appeared on the Lakes during that time. This point is fully discussed in Part I, and illustrated by explanatory maps, showing the mutual repulsion of dissimilar barometers.

On the 19th, a Vulcanian, and on the 20th, a Venusian equinox occurred. Our theory suggests a violent commotion on the Lakes at this time; and most probably it would have occurred had not the high barometer been arrested and driven back by the hurricane approaching the coast of Nova Scotia, because a storm centre had developed on the 18th in the Upper Missouri, and was rapidly progressing eastward when, on the 19th, it was arrested in North-western Iowa, and driven back to its source by the high barometer in Canada.* It did not return until the 20th, and then was deflected by the Canadian high barometer northeastward towards Hudson's Bay.

In the Mount Mitchell, North Carolina, observations, I find the following: (*Ibidem* p. 783–84) "August 18th, foggy with heavy rain from 1.35 A. M. to 8 P. M.; light rain at 11 P. M.; heavy rain during a thunderstorm. August 19th, fair till 7 A. M.; cloudy till noon; foggy and heavy rain from 6.38 to 7.56 P. M. August 20th, heavy rain in the morning; generally foggy, with very heavy rain till midnight. Rainfall during the day 2.08 inches." At Knoxville, Tenn., (*Ibidem* p. 910) August 18th, heavy rain with velocity of wind 20 miles an hour.

---

*Note.—This was storm centre No. 8, of Weather Map for August, 1873. It reappeared as No. 9, on the 20th. See Monthly Weather Review of August, 1873, and also Signal Office Report, 1873.

Such was the weather in the Southern States, while a high barometer rested on Northern New York and Canada, and the terrific tropical cyclone was on its way to the coast of Maine and Nova Scotia. The *Crest of the Wave* had met this cyclone, as already stated within the Tropics, far east of the Windward Islands, on the 13th, the day of Mercury's equinox. At the time of the Vulcanian and Venusian equinoxes, it was passing by the usual parabolic movement northwestwardly between the Bermudas and the coast of Georgia. It now acquired terrific energy, and swept the ocean clean in its path. Its destructive career afterwards has already been described.

According to principles shown in Part I, where the attraction of *similar* and repulsion of *dissimilar* barometers was demonstrated, the great tropical disturbance attracted and absorbed in its cyclonal vortex all the local minor disturbances within thousands of miles of its path. Hence the local disturbances of the Southern States were attracted and swallowed up by it. Thus, little by little, it gathered strength on its way, and amassed and organized as it bore down upon the devoted and doomed coast, that terrible energy with which it smote it upon its arrival. Ah! but, says the tyro in Science, that coast is in possession of and defended by a high barometer! At any other point, and under ordinary circumstances, this would be true, but this is never true of Nova Scotia when a cyclone is coming down the Gulf Stream, for there is a permanent high barometer at or near the Bermudas, and when there is a high barometer in Canada, there is not sea room enough for the cyclone to pass without a terrible and protracted conflict with the opposing forces.

It is therefore a fatal mistake to deduce from general principles that a high barometer gives immunity from cyclones in Nova Scotia or on the coast of Florida, because it does so at points where there is ample room for the storm centre to sweep around, as it invariably does, an interposed area of high barometer.

As early as the 18th of August, the conflict between the tropical low and the Canadian high barometer, commenced, and the latter was driven back northwestwardly towards Hudson's Bay, for reinforcements. It reappeared in Ontario on the 20th, and renewed the conflict, retarding the advance of the tropical cyclone, which, as the map of the Signal Office shows, made very little progress during these days, but the retardation was mutual, for the high barometer also shows extremely tardy movement during this time. The latter is again swung around in a large circle northward on the 21st, but returns, and by a curve towards the northeast, clears the track of the cyclone on the 22d. But on this very day another high barometer appears north of Lake Superior. It advances and encounters the tropical hurricane, now opposite the mouth of Chesapeake Bay; it is overpowered and driven towards Labrador. On the 25th it is at a point to the west of Quebec, whence it is driven northwestward towards Hudson's Bay; but reappeared north of Lake Superior on the 26th, after the cyclone had left for the coast of Scotland. This is illustrated by Maps Nos. 6 and 7, Part I.

In the first part of this work we have shown that the terrific energy of cyclones is generated when a cyclone is compelled to force a passage between two high barometers, between which it is driven like a wedge. Its flanks are driven in upon its centre by the two flanking high barometers. This is the cause why it then acquires that terrific energy that appals and strikes with consternation and amazement every beholder. This Nova Scotia cyclone and the Florida cyclone of October 3d–6th, 1873, both fearfully illustrate this principle.

Moreover, in Part I, we have shown, that whenever there is such a tremendous upheaval, or rather sucking up of air as there is in the bore of a hurricane, there must be an equivalent downpour of air, somewhere within limits, to supply it. We there demonstrated that a high barometer is caused by a down-pour and a low barometer by an up-pour of air. Hence a hurricane which is nothing more nor less than the point of lowest depression within an area of low barometer, requires not only a neighboring high barometer, but an extraordinary high one to supply it with air. This is not only a logical deduction, but a necessity, for which Nature has provided the means. The motive power putting the up-pouring and down-pouring columns of air in motion, we there demonstrate to be Electricity. Hence a low barometer, which is an ascending current of Electricity, will like any other electric current, evoke by Induction an electric current flowing in the opposite direction. That is, a low barometer will evoke or beget a high barometer, and *vice versa*.

We have said that this sad and terrible calamity occurred because of a culpable ignorance of the first principles of Meteorology, which prevents us from foreseeing such dangers, and consequently of taking measures to meet them and avoid the destruction they threaten.

Our Age is laboring under the fatal delusion that all that can be known of meteorological causes, producing atmospheric changes and movements, was known to the good, old, learned, and revered fathers of Philosophy, whose crude notions we have inherited, but not their mental force and vigor. This delusion exacts us to believe just as they believed; to see things just as they saw them; to make observations just as they made them; to pursue investigations just as they pursued them; to think just as they thought, and under no circumstances whatever to arrive at any conclusion different from theirs. But as they were as blind to the Future as bats are in daytime, they had to take events as they came; and we, as long as we adhere to their dogmas, have to do the same. Practically, this is Fatalism in its worst form; for if Nature has deprived us both of the means and the faculty of foreknowledge, then all we can do is to resign ourselves to any fate that may betide.

When we look at the astronomical condition existing at the time the Nova Scotia cyclone supervened, such an exposure of life and property as was there and then, seems like madness; and it seems not only like tempting, but defying Fate.

Two planetary equinoxes had just occurred on two successive days. An equinox of neither of these planets ever occurs singly without manifesting more or less violence, yet now, since they had joined forces, what might not have been expected? From these astronomical events a storm was impending and inevitable; and here was Nova Scotia, situated right in the gate of the highway, not only of all the continental storms of North America, but also of that of the terrific tropical storms that originate in the tropical seas east of the Windward Islands, and by a parabolic movement westward, sweep around the area of high barometer that permanently covers the Bermudas and Sargasso Sea. Whatever other points enjoyed immunity from such visitations, this coast enjoyed none; for here between the mouth of the St. Lawrence and the centre of the Gulf Stream, is the highway of all North American storms; the gate through which they must all pass, even if they have to force a passage towards the attracting permanent Icelandic low barometer in the northeast. Yet under such physical conditions, and in such a locality, we find life and property as recklessly exposed as though hurricanes and their devastations were unknown phenomena upon that coast.

That cyclonal phenomena depend upon and are caused by astronomical events, was then unknown to the public. But a cause with energy adequate to produce such a terrific phenomenon as that cyclone was, cannot conceal itself, nor be without a witness of its presence. The Canadian high barometer—to those who can read correctly the meaning of the barometer—advertised the fact that the impending storm would not be a continental, but a tropical one; and hence would be somewhat tardy in its appearance. The Canadian high barometer, by its slow movement, and in the direction it moved, also gave notice that its way to join the Bermuda high barometer was cut off by the cyclone, and there would be a terrific conflict when the latter would come between it and the Bermuda high barometer. Such was the lesson inculcated by the barometer. But there was still another witness. Under such peculiar and intense electric conditions, the sky and the clouds assume a hue and form so entirely distinct and characteristic as to be unmistakable. If only never so small a speck of sky can be seen, it will be perceived to be intensely serene and extraordinarily blue; and the clouds by their massiveness, compactness and sharpness of outline, either against the sky or against their different strata, or volumes of the same stratum, will give unmistakable evidence not only of the imminence but of the energy of the approaching danger. Even when astronomical events have occurred, or are about occurring, that will bring on paroxysms, the interpretation of barometrical readings, and especially of what is written on the clouds and sky, cannot be dispensed with; for they inform us of the development and progress of the impending phenomenon, its locality, what direction it is taking, and the very moment it may be expected to be upon us if we are in its path.

We find the material we have on hand too vast to be embodied in these pages, and hence will use only so much as is necessary to establish the Planetary Equinoctial Theory beyond gainsaying. Before proceeding to do so, however, we will remark that, so far we have only examined and mainly presented European phenomena, but now we will give attention to American ones, since they serve our purpose better for illustration than European do.

As before remarked, European storms have generally to go too far from home before they fall under observation, to coincide sharply with the date of their exciting causes. The same remark applies to tropical hurricanes, both American and Asiatic. They too, originate out on the ocean, at far distant points not frequented by ships; and hence may exist for a week or ten days before they strike a neighboring coast, or fall under observation of a passing ship. Hence with them also no sharp coincidence will be found when the date of their appearance is compared with the date of the exciting planetary equinox. But with the continental storms of America it is otherwise. They are so near at home that they announce themselves generally a day or so in advance of their planetary equinox. Generally they originate in Western Manitoba (pronounced Man-ec-to-baw), where there is practically a low barometer all Summer. They arrive at the Great Lakes and in the trough of the Mississippi Valley, about the date of their exciting planetary equinox; and on the Atlantic coast a day or two later. In applying the principle of the Planetary Equinoctial Theory to Southern storms, a distinction must be made between the tropical hurricanes that come from the ocean east of the Windward Islands, and those that originate in the Gulf and in the western part of the Caribbean Sea. The former will almost uniformly be found to be from eight to ten days tardy; while the latter will be at most from one to two days only, and often sharply up, if not ahead of time, like their brothers of the northwest are.

The general principle of our theory must already be sufficiently understood so as to need no further illustration. As we find it more convenient to do at once what has to be done, at any rate before the verification and demonstration is complete, so we will proceed and demonstrate that when the Earth is under the influence of the electric excitement produced by either its own equinox or that of another planet, no violent paroxysms ensue, unless by the intervention of the equinox of a second planet. For instance, Jupiter at his equinoxes imposes an intensely high electric condition upon the Earth for a period of three years, yet violent paroxysms only occur when another planetary equinox supervenes.

It is at the Terrestrial equinoxes, during a Jovial term, that the most terrible earthquakes and the most destructive cyclones occur. All the Venusian equinoxes during that period produce phenomena of unusual energy and violence. A Mercurial equinox that ordinarily passes with phenomena so mild as scarcely to attract attention, now gives rise to phenomena of a very violent character. But especially is this the case with Vulcan; strong at all times, his energy then is truly terrific, as we will show at the proper time. As it is during a Jovial period, so it is during a Telluric and Venusian. The violent paroxysms are always brought about by the supervention of the equinoxes of other planets with periods so short that almost the paroxysm can be foretold to the very day. Mercurial and Vulcanian phenomena, especially the latter, are of this character. When we discuss Vulcanian phenomena, we will demonstrate that this especially is the case when only that planet and the Earth are concerned in the phenomena; and in some cases the phenomena seem to be purely Vulcanian when no other disturbing cause, in our present state of knowledge, can be assigned than that of this planet.

We will now take up the Venusian period and verify the principles we have laid down, namely, that the most violent phenomena during that period, nearly always occur at or near the date of a Mercurial or Vulcanian equinox. Our demonstration will make evident, what however might have been inferred, that the crisis in many cases is precipitated and occurs a day or two before the exciting equinox. The Vulcanian equinox of June 17th, 1875, just occurred at this writing, is a case in point. The unusual and extraordinary rainfalls in California, and the tornado at Quincy, Illinois, and other points West, occurred on the 15th of June, two days before the equinox. The terrible rain and wind storm, accompanied by a slight earthquake here, and quite a severe one in Indiana and Ohio, came on the day of the equinox.

## GENERAL PRINCIPLES VERIFIED.

On the 23d of December, 1853, the fine new steamship *San Francisco*, sailed from New York with a regiment of United States troops on board, bound around Cape Horn for California. On the 24th she encountered a hurricane on the Gulf Stream, that in a few moments made a complete wreck of the ship. Her decks were swept, and by a single blow of those terrible seas that storms raise there, one hundred and seventy-nine souls, officers and soldiers, were washed overboard and drowned. She was seen on the 25th by one vessel, and on the 26th by another, but neither, in the tempestuous state of the weather, could render her any assistance.

Capt. Linnell, of the *Eagle Wing*, from Boston to San Francisco, encountered this same hurricane on the 24th. An abstract of his log reads as follows: "December 24th, 1853, first part threatening weather; shortened sail; at 4 P. M. close-reefed the top sails, and furled the courses; at 8 P. M. took in fore and mizzen-top sails; hove to under closed reefed main topsail and spencer, the ship lying with her lee rail under water, nearly on her beams ends; at 1.30 A. M., 25th, the fore and main top gallant masts went over the side, it blowing a perfect hurricane; at 8 A. M. it moderated; a sea took away jib-boom and bow-sprit cap. In my thirty years experience at sea, I have never seen a typhoon or hurricane so severe. Lost two men overboard, saved one; stove sky-light, broke barometer, etc."

This evidently was a Gulf cyclone, and had originated perhaps three or four days before it encountered the doomed ship. The course that storms originating in the Gulf of Mexico will take northeastward, toward the Icelandic low barometer, depends upon the position of the Bermuda, or more properly, Sargasso Sea permanent high barometer. Besides being pushed North, and dragged South by the advance or retreat of the Sun, it constantly oscillates like a slow pendulum, swinging East until it covers the Mediterranean Sea and Africa, as far East as the Nile, when it slowly returns, and in its western oscillation, covers the Gulf and the Gulf States as far as the Rio Grande. When it is on its extreme eastern oscillation, storms originating on the Gulf or in the Gulf States, sweep northeastward along the Gulf Stream. When it is in medium position, such storms are repulsed and driven upon the continent, and sweep down along the sea board; but when it is on a western swing, these storms strike the coast of Texas, and then proceed northeastward, by the Great Lakes, taking Texas, Western Louisiana, Arkansas, Missouri, etc., in their route.* In the case of the storm that struck the *San Francisco*, the high barometer must have been on its eastern swing; hence the storm pursued its natural course, the Gulf Stream northeastward. Had the Signal Service then been established, it no doubt would have detected this storm in the Gulf or on the southern coast of Texas, at or before the departure of the ill-fated steamer, and have avoided the disaster. The astronomical conditions were extremely threatening at this period, and if known, would have shown that it would be the height of imprudence to venture out upon the Gulf stream at the time the ship left port.

Venus had passed her equinox nine days before, namely, on the 12th of December. Vulcan—the Terrible—passed his equinox on the 21st, the day the steamer left; and Mercury would make his equinox on the 26th, or in five days. Under such conditions it would have been a miracle if a cyclone had not passed down the Gulf stream within five or six days from the time the steamer left.

Though out of place here, because Venus was not concerned in it, to enforce the lesson inculcated with such awful sanction by this terrible disaster, we cannot refrain from citing another to the same import, namely, that it is extremely hazardous to venture out upon the Gulf stream, when *Vulcan has just passed, or is about to pass an equinox*, while the excitement of a Telluric or Venusian equinox is impending. The case is that of the steamer President, which left the port of New York on March 11th, 1841; encountered the terrific cyclone of the 13th, and was never heard of afterwards. She had on board a full load of passengers, many of high rank, wealth and influence, both American and foreigners. Amongst the latter was Tyrone Power, the comedian; a son of the Duke of Richmond, etc. This vessel left port while both a Telluric and Mercurial equinox was im-

---

*Note.—My friend, the Hon. Thomas Allen, assures me that these southwest hurricanes frequently do immense damage to his railroads in Arkansas and Southern Missouri.

pending, and both to occur on the same day, in ten days. Vulcan had passed his equinox on the 8th, three days before, not giving time enough for the fierce storms this planet always generates *when one of his equinoxes occurs under such conditions.* This again was a southern storm from the Gulf, and must have been pursuing its destructive career fully five if not more days before it struck the doomed vessel with its living freight, and

> "Sent it to ocean's depths with bubbling groan,
> Without a grave, unknelled, uncoffined and unknown."

A furious hurricane raged on the Great Lakes and in Canada, on the 9th and 10th of October, 1844. Many vessels were wrecked and many houses blown down. Telluric equinox Sep. 21st. Venusian September 23d, and Vulcanian October 9th, day of the commencement of hurricane.

In an English list of wrecks we find: "Screw-steamer *Royal Charter* totally wrecked off Moelfra, on the Anglesea coast; 446 lives lost. The vessel contained gold amounting in value to between £700,000 and £800,000, on night of October 25th and 26th, 1859." As the record did not state whether the disaster was caused by a storm, or was the result of accident, on referring to a list of storms we found: "Dreadful storm on night of October 25th and 26th, 1859; the *Royal Charter* totally lost, and many other vessels; another terrible storm on October 31st, and November 1st, 1859." As the major Jovial equinox occurred early in December, 1859, the Earth was hence laboring under its perturbation. A Venusian equinox occurred October 18th, 1859, or seven days before the first storm. A Vulcanian equinox occurred on October 31st, or six days after the first storm, and on the very day of the last storm. The combination hence was, Jovial, Venusian and Vulcanian. It may here be stated that Mercury passed his equinox November 20th, 1859. On the English list of wrecks I find: "Mail steamer *Indian* wrecked in a storm off the coast of New Foundland; out of 116, 27 lives lost, November 21st, 1859," that is the day after the equinox. The combination here was Jovial, Mercurial and Vulcanian, since Vulcan again passed an equinox on the 23d of November. On a skeleton list of American storms, I find, "Severe storms on the Lakes, November 25th, 1859," no other particulars given.

Prof. Bache, Hare and Beck described a tornado that occurred on June 19th, 1835, in New Jersey, traveling northeastward through New Brunswick, Piscataway, and Perth Amboy, to the ocean. It had at times a double cone, one inverted with base on the cloud, and the other resting upon the Earth. Sometimes it was only a cone or funnel depending from the clouds, changing its position rapidly. It prostrated everything in its path, trees, fences, buildings, etc., and drank up the Raritan river to its bed. Tornadoes are also recorded on this day at Kinderhook, White Plains, etc., N. Y. This tornado occurred about six months before the major Jovial equinox which occurred about January 1st, 1836; and consequently it was partially owing to the Jovial perturbation. An equinox of Venus occurred June 28th, or nine days after the tornado, and both a Mercurial and a Vulcanian equinox had occurred on June 12th, or seven days prior to the tornado.

On 30th of April, 1852, occurred a most remarkable tornado at New Harmony, described by Chappelsmith. It was traced from near Paducah, Kentucky, northeast to New Harmony, and thence 200 miles east to Georgetown, Kentucky. It occurred at 4.30 P. M. This is always the most critical time of day, when if there be any cyclonic tendency, it will surely develope either into tornadoes or into violent wind and hail storms. The same hour in the morning is also quite critical, but less dangerous. If attention is directed to the time of day that tornadoes and hailstorms take place they will almost invariably be found to occur in the afternoon, generally between three and five o'clock. Sometimes they occur later in the day, but seldom earlier. The reason for this is obvious. The storm centre always is the lowest point of an area of low barometer moving forward over the surface of the Earth. Now, since there are two barometrical depressions daily; one, the lesser, between three and four A.M., and the other, the greater, between the same hours in the afternoon, hence, whenever and wherever the low barometer of the moving storm centre coincides with either daily depression, there is danger that either a tornado or violent wind and hail storm may burst upon the locality unfortunately so situated. The day that the tornado occurred was cloudy and threatening, and the barometer quite low. It is remarkable that at four or five miles distant no wind nor any other unusual agitation was observed.

Near New Harmony the track of fallen trees was half a mile in width, and its forward movement estimated at the rate of sixty miles an hour. The destruction was the work of a moment, and intense electric energy was apparent. One observer said, "the cloud appeared on fire at the bottom like a large pile of burning brush." Others say it was "a cloud with green and red flame," and others said it was green and blue.

The following were the impending planetary equinoxes: Venus May 30th; Mercury April 21st, or nine days before the tornado; Vulcan May 1st, or the day after it.

The following, though containing nothing extraordinary, will be read with interest since it is the first case wherein our theory is brought to the test of the actual observations of the Signal Service.

In November and the early part of December, 1870, the astronomical condition was as follows: Equinox of Mercury occurred Nov. 4th; Vulcan Nov. 10th; Venus Nov. 14th; and Vulcan again Dec. 3d. The October equinox of Vulcan, namely, on the 18th, had been signalized by tornadoes; namely, on the 15th at Milwaukee, and on the same day terrific tornadoes occurred in southern Ohio and northern Kentucky. On the afternoon of October 21st, a fearful tornado struck Belleville, Richland Co., Ohio. But we intend to confine our remarks to the earliest observations of the Signal Service, which about the first of November established its stations in the West.

The equinox of Vulcan on the 10th of November, would make us anticipate a disturbance under the conditions so near to Venus, and so soon after Mercury, a day or so in advance of his time. The first dispatch sent from the office at Chicago, is dated at noon on Nov. 8th, and was sent by Gen'l Myer, Chief Signal Officer, and directed to be bulletined by the lake observers at once. It was as follows: "High winds all day yesterday at Cheyenne and Omaha; a very high wind at Omaha this morning; barometer falling, with high wind at Chicago, Milwaukee, etc.; high winds probably along all the lakes." The next morning reports showed that, as anticipated, high winds had fallen on all the lakes.

The next dispatch was sent Nov. 9th, at 11 A. M., and directed to be bulletined on the lower lakes, at New York, and at

Boston: "Low barometer, moving eastward; high winds along the lakes, and probably will be along New York and eastern coast." This again was verified.

On the 19th, another disturbance appeared in the northwest, and signals were ordered to be hoisted, but it seems it passed northward of the lakes, manifesting itself feebly only at Duluth and Detroit. Nothing serious was now to be expected until the recurrence of a Vulcanian equinox on Dec. 3d. Here is what a paper published in the Annual Report of the Chief Signal Officer for 1871, says: "The first considerable storm that swept over the lake region after the commencement of these (Chicago) reports, was that of December 6th–7th, 1870." (p. 168.) The origin and progress of this storm is minutely described and illustrated by twelve maps, but we have not room for details. The following must suffice:

"Dec. 3d, 4 A. M., the following dispatch was sent: Barometer rapidly falling at Omaha, St. Paul and Duluth; heavy weather probably on the lakes. It was received at Duluth, on the western extremity of Lake Superior, on the day of its date, and within one hour afterward the storm set in with great violence. Wind from the northeast, and the lake very rough."

As the high barometer to the east and the southeast, shown on the maps, repelled the low barometer or storm centre, it remained almost stationary in Kansas and Nebraska, until the high barometer fell back slowly on the 5th. At 10 A. M. this day, the following dispatch was sent to all the stations on the lower lakes: " Very low barometer, with rain at Milwaukee and Chicago, progressing eastward." "The *Evening News*, published at Cleveland, Ohio, published this dispatch, and the *Ledger*, next morning said, 'At 3.20 P. M., the storm arrived, falling at once upon the city in the utmost fury, drenching the streets with floods of water, wrenching off signs, knocking down chimneys, and causing the wildest consternation among pedestrians, who were caught without a moment's warning in a terrific gale.— (*ib.* page 170.)

"On the 6th," we are told, "the very low barometer still lingered on Lake Ontario, where rain continued to fall, which extended south to Pittsburgh," supplemented with this speculation: "Perhaps this retardation of the storm may have been occasioned by the highlands of the Adirondacks, and those of the Eastern States." Why then do not these mountains retard every storm? Or, are they cause only at times, and no cause at other times? We would also like to have the "highlands" pointed out to us that checked the progress of this low barometer down the valley of the Missouri, kept it stationary for nearly two days at Omaha, then suddenly deflected it on the western point of Lake Superior, and afterwards retarded its progress over the upper Lakes. When that is done we will be able to appreciate the theory that the Adirondacks, the Green, and the White Mountains retarded its progress and made it linger so long on Lake Ontario. We have given as an explanation of the retarding and deflecting phenomena of storms, the mutual repulsion of high and low barometers; and until a case is shown where a low barometer has pushed right into the centre of a high barometer, or *vice versa*, or that either has kept on its course indifferent to, or regardless of the presence of the other, we shall adhere to that explanation. It is a mere question of fact, and a five minutes examination will settle it. The maps themselves, illustrating this storm, establish the correctness of our explanation beyond gainsaying.

At the conclusion of the earthquake testimony in relation to the Jovial Cycle, we stated that we would produce additional testimony on that point, when we came to verify the Venusian Cycle. We will now redeem that pledge; not as cumulative proof of the reality of that Cycle, for that we consider incontestible, but for the purpose of demonstrating how phenomena crowd, whenever cycles intersect each other.

The Jovial Cycle culminated as we have seen at the Autumnal Equinox of the year 1871, the phenomena of which have already been given. We will now call attention to the phenomena of the Vernal Equinox of that year; when we have superimposed upon the Jovial perturbation the following accessory disturbances, namely, that of the Earth's equinox on the 21st of March, that of Venus on March 5th, that of Mercury, March 16th, and those of Vulcan February 10th and March 5th.

Auroras were observed from the 9th of February to the end of the month, on every day excepting on 14th and 25th. Those of the 10th, 11th and 12th very brilliant; those of the 10th and 11th extraordinarily so. Magnetic, that is, electric disturbances, were observed in Europe on the 12th, and an aurora was seen by daylight in England on that day. Electric disturbances were observed in the Southern Hemisphere on February 4th, 5th, 9th, 13th, 14th, and 15th. February 9th,—one day before Vulcan's equinox,—an earthquake occurred at Illapel, Chili, simultaneous with an extraordinary rain, causing destructive freshests from the Andes. On the 1cth, tremendous rains in the Andes of Peru, causing destructive floods. February 11th— A slight shock of an earthquake at Valparaiso, Chili, simultaneous with another tremendous rain in the mountains, causing disastrous freshets. In Hayti, on the 17th and 19th, severe shocks of an earthquake were felt. Severe earthquakes were felt on the 18th in Burmah, and westward as far as Calcutta. On the 19th, a terrific earthquake occurred in the Hawaian Islands. Earthquakes were also felt in Peru on the 22d and 23d, and on 25th a very severe one in Chili. On the 25th, also a violent earthquake occurred in Comiguin, one of the Philippine Islands, which continued to May 1st, when it culminated in a terrible disaster* at the Mercurial equinox.

March 2d—An earthquake in Nevada; same day detonations commenced in the volcano Roeang, Celebes; on the 5th, at 7 P. M., a frightful eruption took place; three minutes later a wave reached the shore of Tagoeland, one mile distant, destroying three villages and drowning 416 persons. The eruption continued till the 14th, when the heaviest and final eruption took place. March 6th—A severe earthquake at Bogota, commencing on the 4th.

March 5th—Earthquakes at Arequipa, Jacua, and other places in Peru, said to have been preceded by an electric storm, and accompanied with tremendous rains, followed by terrible floods. Near Lima, many lives were lost, plantations wasted, and railroads washed away. March 6th—A most remarkable electric storm at Tacua, Peru; for hours the snowy peak of Tacora seemed the centre of a conflagration of lightning and terrific thun-

---

*Note.—The following is the newspaper account of this catastrophe: A terrible earthquake in the Philippine Islands. On the island of Comiguin, after being terribly shaken since February, and especially during March, a plain fell in on the 1st of May, 1,500 feet in diameter, engulphing many houses, cattle, and over 150 persons. The gulf became a crater, and is still casting up ashes, stones, smoke, etc.

der, that shook the mountain and the country for miles around. On the 6th, an earthquake was also felt in New Hampshire.

March 8th—The terrific tornado of East St. Louis occurred, fully described and discussed in Part I. March 17th—Another remarkable electric storm in Peru, and very brilliant auroras seen in both hemispheres. In England an earthquake and remarkable earth currents in Atlantic cables. In Scotland, the earth currents were immediately followed with a remarkable rise in temperature, from 17° to 92° in fifteen hours. On the 20th, another shock of an earthquake occurred in England.

As a Jovial equinox was impending, it could be anticipated that the phenomena attending the June, 1871, equinox of Venus, would not only be frequent, but that they would exhibit extraordinary energy. The disposition of the equinoxes not only ensured frequent paroxysms, but a protracted period of them. The following was their astronomical arrangement: (1) An approaching Jovial equinox, occurring September 25th. (2) Vulcanian equinoxes, occurring June 5th and 28th. (3) A Mercurial equinox, June 12th; and, (4) the Venusian, on June 25th. The following is an abstract of my phenomenal record for the month of June, 1871:

June 1, a terrific rain fell in the great rainless desert of Atacama, Chili, on May 31st and June 1st.

June 2d—Tremendous thunderstorm at this point (St. Louis) and vicinity this afternoon; at the same time a terrific tornado occurred in Macon county, Ills.* 5th—A heavy thunder and rain storm, coming from the western part of the Gulf, passed over the Indian Territory, Missouri, etc., towards the Lakes, followed by showery weather for three or four days. 8th—Earthquake at Waggawagga, Australia. 10th—A brilliant aurora seen by me at Golden, Colorado. 11th—Another heavy thunder and rainstorm from the Gulf passed northeastward over Arkansas, Missouri and Illinois, to the Lakes. 12th—A terrific hurrican at Galveston, Texas. 16th—An awful tornado at Eldorado, Kansas, the whole town was destroyed; on the same day a terrific hurricane in Louisiana. 17th—Witnessed a most brilliant aurora at Denver, Colorado. 18th—Encountered an intensely

---

*Note.—This tornado is fully described in Part I.

hot simoon on the Plains, between Wallace and Bosland. On the 17th and 18th—a most remarkable magnetic disturbance was observed at the Observatory of Havana, Cuba, commencing at 10 P. M., and continued 24 hours. 18th—The following phenomena are recorded: Violent tornadoes at several points in Wisconsin; a terrific tornado at Scranton, Green Co., Iowa, demolishing houses, and carrying dwellings ten rods and killing the inmates; a terrible tornado at Westerville, Iowa. An awful cyclone several miles east of Springfield, Illinois, accompanied by a terrible roar and a bluish flame reaching from the Earth to the clouds; it pulled up trees, and gathered up fences and everything in its path, and whirled them in the air. A very brilliant aurora observed in Europe, accompanied by violent magnetic disturbances, and synchronous with immense storms of rain in England, France, etc. Earthquake in New Jersey and on Staten and Long Islands. In New Jersey the earth opened and swallowed up trees; a gulf was sprung in the canal, through which all the water disappeared: it was found a tedious, difficult and costly job to fill up and close the rent. 19th—Terrific thunder and rainstorm, flooding the country in eastern Kansas and Missouri; it was accompanied by a heavy gale. Same day an earthquake in Brooklyn, New York and New Jersey, at 10 P. M., accompanied by a rumbling sound. 29th—Earthquakes in the island of Madeira, in Chili, and in Peru; very violent at Tacua. 21st—An earthquake at Calistoga, California; and a cyclone at Surat, India, devastating the cotton crop. 26th—Earthquake at Chiriqui, Central America. 28th—A violent earthquake at Tocray, Peru; same day occurred the remarkable tornado in Ulster County, New York, described in Part I. On the night of 27th-28th—A most remarkable storm on Lake Superior, followed by a destructive tidal wave at Duluth. The storm must have been accompanied by a tremendous water-spout. It is described as follows: "An awful rainstorm, with vivid lightning but very little thunder, was over the Lake all night. The lightning flashed up in sheets from the Lake; the rain first fell perpendicularly, then came furious winds from all points of the compass, that lashed the water into mountain waves, and twirled them into spires, till they reached the clouds."

The phenomena of July belonging to this Venusian disturbance are as follows, but we will only quote the first nine days:

July 3d and 4th—A faint diffused aurora, as seen from St. Louis along the whole northern horizon; during this night and morning a general rainstorm prevailed from Wisconsin west to Wyoming. The cloud was so intensely electric as to make telegraphing impossible; and in many places there were terrific winds accompanied with destructive hail. 4th—The disc of the Sun covered with spots; a violent typhoon at Kobe, Japan; great damage at sea, many vessels wrecked, and on land many houses blown down, and over 400 persons killed. 5th—A destructive typhoon at Hioga, Japan; an earthquake in California; dreadful tornadoes and water-spouts at Rowen, Reno, and Truckee, Nevada, doing immense damage to railroad; terrible tornadoes in Eastern Nebraska and Western Iowa; a train blown off the track and one car carried over 200 feet, at DeSoto, Nebraska. 6th—A tornado at Mandeville, West Virginia. 7th—A destructive tornado in Arkansas. 9th—A water-spout in Cork harbor, Ireland. At 4 P. M. a violent and destructive tornado occurred at Dayton, Ohio. Many houses were demolished and many churches and bridges blown down. Many persons were killed, and the damage done to property in the city and county, estimated at over a million of dollars.

The phenomena of the succeeding Venusian equinox in October, 1871, have mostly been given when the Jovial Cycle was under discussion, therefore it is not necessary to quote what remains of them, nor to quote those of succeeding cycles. We indeed might take up every Venusian Cycle that has occurred since, and produce incontestible evidence that similar phenomena have characterized each of them. We could do more; we could go back indefinitely and do the same, as we have ample material on hand. But this would not only be tedious, but superfluous. If the evidence we have produced from the records of the Past; and if the testimony that Nature has furnished in the phenomena she has exhibited since we first published our theory to the world, challenging a test of it by facts; and especially, if the astounding phenomena daily occurring while we are writing this (June, 1875), do not convince every one of ordinary intelligence of the reality of the Venusian Cycle, then he is too incorrigibly dull to be convinced by any evidence whatever, or by facts even that would lay waste and desolate whole continents. Before closing the discussion on this point we will add, since, to give the facts themselves, would be a mere repetition; for by changing the names of the localities, the detailed account we have just given of the phenomena in June, 1871, is exactly descriptive of the phenomena of June, 1875. To us the coincidence and identity of phenomena of such extraordinary character, happening after so long an interval of time, but under identical astronomical conditions, is conclusive proof that those conditions not only influenced but caused the phenomena.

The Terrestrial Cycle, we may say, has always been known; not indeed as a meteorological cycle in the sense we must now understand it; but from time immemorial it has been known that when the Sun apparently, but the Earth really, crosses the Equator, as the Equinoctial Colure is called, storms prevail, which have received the characteristic name of "equinoctial storms." It has also been known that simultaneously with the occurrence of these storms, a meteorological change takes place over the entire Globe; the polar hemispheres exchange seasons and climates; and the direction of the wind over the whole surface of the Earth is more or less changed, and in some localities entirely reversed, in consequence of the hemispheres changing their electric state from the static to the dynamic, and *vice versa*. These great changes are wrought by a change in the electric condition of the Earth, which at one equinox renews its electric energy at the North, and at the other at the South Magnetic Pole of the Sun. From well established electric laws, it might have been foreseen that the Earth, as a member of the Solar System, could not do so without effecting the electric condition of every member of that system, as well as that of the great central Luminary himself.

It is an astronomical fact, that the Earth's orbit is so placed in reference to the Sun, that the Earth in every revolution on opposite points of its orbit, is alternately brought, so to speak, above and below the Sun; and that its equinoxes occur when it passes by the Sun. As all the orbits of the planets are similarly placed, and make various angles with the Earth's orbit, therefore each planet in every revolution passes, in popular language, above and below the Sun; and in passing by the Sun, its equinoxes occur by which its electric condition is affected. But by an inexorable law, this cannot take place in any of them without affecting the electric condition of all members of the system.

We have proven that both when Jupiter and Venus pass these points on their orbits, they effect great changes in the electric condition of the Earth and of its Atmosphere. Hence it is only an extension of the same principle when we logically, as we inevitably must or else be inconsistent, infer the same influence will be felt, and the same effect produced by all the other planets under similar circumstances.

We have already given the accessory agency of Mercury in causing phenomena, pending a Venusian disturbance. It will be seen by reference to the facts stated, that the Mercurial period is always sharply defined, since the phenomena often appear at the exact time that his equinoxes take place, and always within a few days before or after. Being so well defined, the Mercurial Cycle will not be seriously questioned; but determined not to leave any point in doubt, we will endeavor to show his direct agency by adducing cases wherein his influence is not complicated with extraneous influences. This expression, however, must be understood in accordance with the principles already stated; namely, that all phenomena are compound, and that no planet, not even Jupiter, is sole cause of them.

All planets exacerbate the electric condition of the Earth, but as this exacerbation, when caused by slowly moving planets, comes on gradually and abates in the same manner, no electric paroxysm ensues, unless by the supervention of a disturbance of a swiftly moving planet; and then the more suddenly this disturbance supervenes the more well-defined and violent is the paroxysm. For instance, the Jovial perturbation,—if the observations of magnetic disturbances are to be credited,—comes on gradually for two years and upward; but the full force of a Venusian disturbance is enforced in thirty days; a Mercurial one in ten; and a Vulcanian in not exceeding seven days. Hence the more rapid the orbital velocity of a planet, the more complete will be the juncture of its phenomena with its equinoxes. Hence, too, we find that Mercurial and Vulcanian phenomena always,—unless they originate at localities distant from the place of observation,—sharply coincide with their equinoxes. For real Mercurial phenomena we must look to such as occur at periods when there is a Jovial disturbance prevailing, and which are too remote from other planetary equinoxes to be affected by them. This is a somewhat difficult task: since Mercury, from the smallness of his size, were it not for the energy of his position near the Sun, would scarcely exercise any perceptible influence. But as it is, his influence is decidedly more marked than that of Mars.

The following, though it was not a pure Mercurial phenomenon, yet it deserves a record: The *Lady Nugent*, a British troopship, sailed from Madras May 10th, 1854, and foundered in a hurricane a few days afterwards. All on board, 350 rank and file of the Madras infantry, officers, and crew, altogether over 400 perished. A Jovial equinox had taken place about the middle of December, and a Venusian April 4th. A Vulcanian and a Mercurial had occurred on the same day, May 8th, or two days before the vessel sailed. The hurricane occurred in consequence of superimposing the Vulcanian and Mercurial perturbations upon the Jovial.

On the 6th of September, 1838, the steamer *Forfarshire*, from Hull to Dundee, was wrecked in a violent storm near the Outer Fern Lighthouse. Out of fifty-three persons, thirty-eight perished. James Darling, the keeper of the lighthouse, and his heroic daughter, Grace Darling—although there was a tremendous sea running—ventured out in a coble, and rescued 15 persons. Mercury passed his equinox on September 7th, 1838. No other equinoctial excitement except that of the Earth prevailed at the time.

On the 27th of August, 1826, a most extraordinary hurricane occurred in the White Mountains, New Hampshire. It brought down a land slide that buried the Willey family. Mercury passed his equinox on the previous day, namely, the 26th. No other equinoctial influence prevailed excepting that of the approaching Autumnal equinox of the Earth.

My phenomenal record of 1871, when a joint perturbation prevailed, affords several instances of Mercurial disturbances during periods when the Earth was free from any extraneous excitement other than the Jovial. The first is that of January 31st. Brilliant auroras were seen, both in America and Australia, on

January 30th; and violent earth-currents were observed at all physical observatories. On the 31st, severe earthquakes occurred at Bombay and Assam, in India, and in Asia Minor they did much damage. On February 1st, 2d and 3d, a general rain and snow storm prevailed in the United States.

At the time of Mercury's equinox, April 29th, 1871, the following phenomena are recorded:

April 24th—Destructive hailstorms near New Orleans; many plantations ruined, and cotton must be replanted.

April 26th—Destructive hailstorm at Rose Hill, Missouri, at Detroit, Michigan, and at Bluff, Texas. 27th—A destructive hailstorm at Canton and Grenada, Mississippi, at 5 o'clock, P. M., and at Ponchatoula, Louisiana. 28th—A terrible hailstorm at Wytheville, Virginia. 29th—Terrific gale and hailstorm at Jacksonville, Illinois. 30th—A destructive hailstorm passed through northern Mississippi; a tornado and hailstorm at Louisville, Kentucky, and a second hailstorm at Wytheville, Virginia. On the same day there was a violent earthquake at Hayti.

We have already mentioned the falling in of a plain, and the engulfing of houses and their inhabitants, on the island of Comiguin, during a protracted earthquake on May 1st. On the 2d of May there was a terrible hurricane at Baton Rouge and the Lower Mississippi. The remaining equinoxes of Mercury for 1871, are all so complicated with those of other planetary equinoxes, as to make it impossible to show what share he had in the phenomena that occurred at the periods. The phenomena of his equinoxes that fell within the Venusian periods, have already been given.

An uncomplicated equinox of Mercury occurred March 3d, 1872. In the Signal Office Report of 1872, p. 288, a full report, accompanied by seven maps, is given of a storm that originated in the Gulf of Mexico, on the 1st of March, and left the coast of Nova Scotia, where it did much damage, on the night of the 3d. March 3d—A severe earthquake at Mostar, in the Herzegovina, and a gale on Lake Michigan.

The next equinox of Mercury is also free from extraneous influences. It occurred on the 15th of April, 1872. In the disasters of the Great Lake, (see Signal Report of 1872, p. 196,) we find, April 13th—Scow *Annie Comine* struck a pier, and afterwards damaged by drifting against a dock. Scow *Nettie* dismasted in a gale. April 14th—Scow *Christie* sunk. Schooner *Union* lost anchor and damaged. Schooner *Game Cock* damaged. Steamer *Jay Cooke* slightly damaged. *Eva M. Cone* lost foretop. *Minnie Corbett* lost two spars. April 15th—Schooner *Liberty* dashed to pieces on a pier in a violent northeast gale. Propeller *Navarino* sunk. Steamer *Sheboygan* driven back by gale, and twenty other schooners and scows damaged, driven ashore, etc., and fish-boat *Hattie* lost with all on board.

My phenomenal record gives April 14th—A violent typhoon at the Philippine Islands, and April 15th—A terrific eruption of the volcano Merapi, Java. Immense destruction of life and property.

The next Mercurial equinox occurring May 30th, 1872, was complicated with a Venusian that had occurred on the 28th. The list of Lake disasters shows considerable damage to shipping from 26th of May to 2d of June. But as it is impossible to separate the phenomena, being the joint production of the two planets, we will not quote them.

The following is from the daily journal of the Signal observer on Mount Washington, and gives an idea of the condition of the weather: May 27th, 1872—Barometer falling steadily all day. Weather fair till 3 P. M., when dense stratus clouds, covered the summit and remained there until rain began to fall at 11 P. M. This weather continued through the night. May 28th—The weather has been very bad all day. Snow fell during the night, and continued up to 10.30 P. M., changing then to rain and sleet. May 29th—The wind from the north-west, with high and gale velocity. The rain last night changed to snow, and continued to fall to a little before day-break; and some of the drifts measured upward of fifteen inches in depth. Heavy stratus clouds obscured the sky and covered the summit all day. May 30th—Day very cloudy. At 6.40 P. M. a heavy driving rain commenced, which continued when midnight report was sent. Difficulty in telegraphing midnight report to Boston, although a strong current was put on, as the bad weather caused frequent breaks. May 31st—The weather has been one continuous round of snow, hail, sleet and rain all day." *Signal Service Report*, 1872, p. 215.

For the Mercurial equinox of July 13th, 1872, from the indefinite way in which the damages to vessels on the Lakes are stated, it is impossible to say whether it was sustained by gales or not; and as the Signal Office then had not commenced their Weekly and Monthly Review of the Weather, the influence of that equinox upon the weather cannot be shown. However, in the Report of 1873, the observations of Sergeant Fish, made at Unalaska, Alaska, are given. By reference to them I find it had not rained at that station from the 7th of July to the 11th, when a low barometer set in with rain daily till July 17th.

The Mercurial equinox next in time was that on August 26th, 1872. In the Lake disaster list are found the following items:

Schooner *Naragansett* lost canvass in a gale. A scow driven ashore on August 26th. On the 27th, schooner *Luddington* lost jib-boom. 28th—A northeast storm all day on Lake Michigan, driving many vessels into port for shelter; scow *Ida Bloom* struck a pier and lost jib-boom; several small sail boats lost or damaged at Milwaukee; a canal boat sunk and another damaged. Schooner *Louise Meeker*, with 22,000 bushels of oats, struck by a gale, capsized and sunk; schooner *Glad Tidings* lost jib-boom in a gale; scow *Minnie Corlett*, loaded with shingles, damaged. 29th—The following schooners suffered damage: *Garibaldi*, sail torn; *Angeline*, lost mainsail; *Sanburn*, split mainsail and foretop sail; and *Almira*, damaged against piers. Scow *Porter* went ashore; and a raft of logs broke loose from a tug and washed ashore, etc.

Mercury's next equinox occurred on October 9th, 1872. There was no list of Lake disasters published from August 31st, 1872, to January 1st, 1873; consequently we cannot refer to the invariable gales on the Lakes at the planetary equinoctial periods. Referring to the monthly weather map of the Signal Office, for October, 1872, we find that storm centre No. 2 of that month, appeared in Dakota, on the morning of the 8th of October, was central on the Lakes on the 9th, and disappeared in the Gulf of St. Lawrence on the night of the 10th.

Mercury's last equinox in 1872, occurred on the 22d of November. There was a storm centre from the southwest that was central on the Lakes on the 24th. The Weather Review for the month of November, gives no special information; it only says:

" During the month four storm centres crossed the country diagonally from the southwest to New England and the British Provinces." This of course was one of them. It further says: " Destructive gales have attended the progress of many of these storm centres, particularly over the upper lake region, in the St. Lawrence Valley, and on the middle and east Atlantic coast."

The first Mercurial equinox in 1873, occurred January 5th, just three days before the Venusian equinox. Its perturbation is therefore complicated with that of Venus. A storm centre appeared in Wyoming on the 3d, passed through Arkansas on the night of the 4th, up the Ohio Valley on the 5th, and reached Nova Scotia on the 6th. On the latter date, another storm centre appeared in the Upper Missouri Valley, and was central on the upper lakes on the 8th. This storm is the one that was so destructive to stock on the Plains in Kansas and Nebraska.

The second Mercurial equinox occurred on the 18th of February, two days after a Vulcanian equinox. The phenomena, according to the Weather Review of February, 1873, were: A storm centre moved, February 15th, 16th and 17th, northeastward from Texas over the Ohio Valley, then eastward over and beyond the Atlantic coast, accompanied by rain, generally heavy; with a severe thunderstorm at Memphis on the 15th. Another on the 17th and 18th, northwestward from Dakota, over Minnesota and Lake Superior, followed by low temperature. And still another on the 20th and 21st, over the Southern States and the northeast, accompanied by high winds and heavy rains and snows over the whole country.

On the 3d of April, 1873, another Mercurial equinox occurred simultaneously with a Vulcanian. In the Lake disaster list we read. " April 2d—Scow *Raven* dismasted. 3d—Propellor *Fremont* damaged. 4th—Schooner *Caroline Marsh* lost her jib-boom. 5th—Light-house at Erie blown down, and schooner *Aldebaran* lost mast by lightning." In the Monthly Weather Review for April, 1873, we read that a storm centre passed over Missouri, Illinois and Michigan, into Canada, on April 1st and 2d, sending minor disturbances over the middle Atlantic coast, accompanied with brisk to high winds and heavy rains in all the States east of the Rocky Mountains. It was felt as a severe storm from northern Texas to the northwest and Lakes. A second storm centre passed over Nebraska, Kansas, Iowa, Missouri, Illinois and Michigan, into Canada on 3d, 4th, 5th, 6th and 7th of April, with brisk and occasionally high winds, accompanied with rain and snow. It sent out several minor disturbances eastward to the Atlantic coast. A third centre passed from Texas to Canada, by the Lakes, on 7th to 10th, with heavy rains from Texas to Minnesota, and eastward to the Atlantic, followed by a severe "Norther" in Texas.

On 17th of May, another Mercurial equinox occurred, and a Vulcanian on the 19th. On the 15th—Schooner *Mary Battle* lost main and mizzen mast in a storm on Lake Erie. 18th—Several schooners driven ashore; many minor disasters occurred up to and including the 21st, which we must pass over. In the Weather Review for 1873, we find storm centre No. 6 travelling very rapidly from Minnesota to New England on 12th and 13th, accompanied with light rains and high winds throughout its course. On 13th, 14th and 15th—Storm centre No. 7 passed from the Plains of Kansas and the Indian Territory through Alabama to the Atlantic coast. Considerable rain fell in all the Southern States. Storm centre No. 8 passed from the southwest by way of Lakes Michigan and Huron, into Canada, on the 18th, 19th and 20th, with violent rain-belts and from brisk to high winds. Storm centre No. 9 passed, from May 21st to 24th, from Montana to Nova Scotia. It was during this disturbance, May 22d, that the terrific tornado in Washington County, Iowa, and another on same day in Central Illinois, occurred, fully discussed in Part I.

The equinox that occurred next is that on the 30th of June. The list of disasters on the lakes mention damages of such a nature that they likely were caused by gales, but being uncertain, we will not quote them. On the 1st of July, however, we have something definite—Schooner *Minnie Mueller* lost foremast and flying-jib in a gale; bark *Mary Battle* arrived with foretop-mast gone; and bark *Dundee* with sail damaged. Tug *Bartlett* lost smoke-stack. In the Weather Review of June, we have storm centre No. 9 passing from the northwest over the Lake Region, from the 25th to the 28th, sending out branches with severe thunderstorms, from Tennessee to Minnesota; the Lakes and New England. From the 28th to the 30th—Storm centre No. 10 passed from Minnesota to Canada, accompanied with severe thunderstorms, north and west of the Ohio Valley. Brisk winds over the northwest, the Lake Region and the Atlantic States; and rain, often quite heavy, from the Lower Mississippi, to the northwest and Lakes, to the Atlantic coast.

The effects of the Mercurial equinox in August, have already been noted, and the origin of the terrific Nova Scotia Cyclone traced to it.

The list of lake disasters was not continued beyond the 31st of August, 1873, as stated before, therefore we cannot refer to the weather on the Lakes at the occurrence of the Mercurial equinox September 26th, 1873. The Weather Review for September says: "Sept. 23d and 24th, high winds over the northwest and Lakes; rains from Missouri and Ohio Valleys to the Lakes, and Middle and east Atlantic coasts. This was the severest storm of the month, especially on the Upper Lakes, where there were very heavy gales. Sept. 25th and 26th—High winds in the northwest and on the Upper Lakes; with rain in Montana, and snow on the 26th; heavy gales on Upper Lakes, Sept. 27th, 28th and 29th—High winds, on Lake Ontario and in Lower St. Lawrence Valley; rain in all sections east of the Rocky Mountains, followed by a severe 'Norther' in Texas on night of 29th." On the 29th and 30th—A high barometer from the northwest covered the entire country, accompanied with low temperature and heavy frosts in the northern sections. I desire to call particular attention to this fact of high barometers following low barometers or storm centres. All our excessively cold weather in Winter is caused by these high barometers preceding or following low barometers. This we will show at the proper place.

The next Mercurial equinox occurred on November 9th, 1873. The Weather Review of that month says: "No. IV storm centre originated in the western part of the Gulf of Mexico, and moved along the Gulf and the Atlantic coast all the way to Nova Scotia. It started upon this track on the morning of the 6th, and reached Plaister Cove, Nova Scotia, on the morning of the 9th. It was accompanied by fresh winds and rain on the seaboard, and on reaching the coast of Maine it was marked by high and dangerous winds in its front. As usual with cyclones taking this track, as it neared Nova Scotia, the barometric depression increased, and the cyclonic winds became more violent, than when the meteor was moving on lower parallels of latitude. Storm center No. V probably originated in Colorado. It was first observed on the 10th in Kansas, and first moved to St. Paul, Minnesota; thence on 11th towards Milwaukee, sweeping eastward with destructive force over the Lakes, accompanied with heavy snow and rain."

The last Mercurial equinox in 1873 occurred on the 23d of December. The Weather Review for that month says: "No. VIII storm centre began on the 17th in the Southwest, and advanced slowly in a direction almost due north-east over the Ohio Valley. It was attended by heavy precipitation and high winds. It disappeared near Nova Scotia on the 20th. No. IX commenced on the 21st, near the mouth of the Rio Grande, skirting the Gulf coast, it crossed over northern Florida on the night of the 22d; thence making its pathway along the main axis of the Gulf Stream. No. X began in the Gulf on the night of the 24th; passed over Southern Florida and thence along the Atlantic seaboard. It reached the coast of Nova Scotia on the 27th. This gale was followed by heavy snowstorms in New England, the snow falling two feet deep in some places."

The Mercurial equinoxes of the year 1874 down to the present time, July, 1875, we will have to pass over hastily, quoting the substance of the statements of the Monthly Weather Review.

The first occurred February 5th, 1874. A storm centre passed from Southwestern Texas, from the 3d and 4th, to the coast of Virginia on the 7th, accompanied by heavy rain and snow in the Southern, Central, and Middle Atlantic States. On the 7th, a Cuba storm appeared on the coast of southwest Florida, which became a very destructive cyclone on the Gulf Stream opposite the middle Atlantic coast.

The second occurred March 21st. Three minor storm centres passed over the Continent from the Northwest, from the 19th to the 25th. On the 16th and 17th there were forty consecutive hours rain in Georgia, and 5.1 inches of water fell.

The third occurred May 4th. A storm centre from the northwest was central in Kentucky and Tennessee on the 4th. It came from the Pacific Coast, and entered the Atlantic off Cape

Hatteras. A high barometer over the Lakes deflected it and made it take this unusual direction.

The fourth occurred on the 17th of June. A storm centre that was central over Kansas on the 14th, passed over the Lakes to Halifax, where it reached on the 18th. This storm exhibited "the steepest barometric gradient for the month." On the 18th a severe shock of an earthquake was felt at Salt Lake City, Utah.

The fifth occurred July 31st. This equinox is complicated with that of Venus, which occurred eight days before, namely, on the 23d of July, and Vulcan on 30th. This month throughout was marked for its destructive local storms. First, those of the 4th and 7th, that were produced by the Vulcanian equinox of July 7th; and second, those that were the joint production of the three crowding equinoxes of the seven last days of the month. These were waterspouts on the 22d and 23d in Colorado; 24th, destructive storm and waterspout in Nevada; 25th, a destructive waterspout in Germany; and 26th, the terrible storm and waterspout at Pittsburgh, or rather at Allegheny City, on the other side of the river, at which 134 lives were lost, according to the official report, and property valued at over five hundred thousand dollars was destroyed. An earthquake occurred at Cairo, Illinois, on the 9th, referable to the Vulcanian equinox of 7th; another at Camp Russell, at Nebraska, on the 23d, the date of the Venusian equinox; and another on August 3d, at San Bernardino, California, referable to Mercury.

The sixth occurred on September 14th. Three separate storm centres passed over the Continent from the 13th to the 18th, with heavy rains, ending the drought that had prevailed in the Middle States and east Atlantic coast for five or six weeks. On the 16th and 17th much damage was done in Kansas, Iowa and Missouri by a severe wind and rain storm.

The seventh occurred October 27th. A severe storm was central in Nebraska on the 27th. Its passage over the Upper Lakes resulted in numerous disasters. A Detroit paper says: "The gale moved everything on land that was not fastened." There was a severe "Norther" in Texas on the 29th. Snow fell so heavily in Dakota as to obstruct the railroads; and a water-spout occurred on Lake Erie on the 31st, striking the shore half a mile west of Buffalo, where it burst.

The eighth, and last Mercurial Equinox of 1874, occurred on December 10th, 1874. A storm centre passed through the British Possessions, north of the Lakes, on the 10th and 12th, with high winds on the Lakes. Another storm centre appeared in the Northwest on the 12th, and passed over the Lakes on the 13th, accompanied by rain in the south, and sleet, snow and high winds in the Lake Region and eastward, with severe gales on the Middle Atlantic coast; several wrecks occurring north of Cape Hatteras. Distinct shocks of a earthquake occurred on the 10th on Long Island, in Westchester and Rockland counties, along the Palisades and Hudson River, New York, and in New Jersey. On the 12th, another shock was felt at Garrison's, New York, and on the 8th there was an earthquake at Guadaloupe, West Indies.

In the present year (1875) the Mercurial equinoxes, up to the time of writing this, have been as follows:

First, on the 23d of January. A storm centre passed from Portland, Oregon, to Canada, from 19th to 22d, accompanied with heavy snows, there were light snows and rains in the Southern States and a "Norther" in Texas. On the 21st to 25th another storm centre passed from the Pacific Coast to Canada, accompanied with severe gales, rain, sleet, and snow, North; and heavy rains and thunderstorms South. At 4 A. M. on the 24th, two shocks of an earthquake were felt at Carson City, Nevada. The first shock was light, the second quite severe, and lasted several seconds. The same shocks were felt at Sacramento, Cal.

The second occurred on March 8th, but is so much complicated with the Venusian equinox of the 5th, and other equinoxes, that its distinct phenomena cannot be traced. On the 6th a storm centre appeared on the Gulf coast of Louisiana, preceded by rains and easterly winds in Alabama and Mississippi. Its path was almost due northeast through Alabama, East Tennessee, thence skirting the Atlantic coast to Halifax. In the Southern States immense rainfall, and in the Northern snow, along its whole route. At Memphis the deepest snow ever known fell. Through the Ohio Valley, thence towards Lake Erie, the snow fall was extraordinary. High winds in the West; severe gales along the Atlantic coast, and on the coast of Texas a severe "Norther" prevailed for twelve hours.

The third Mercurial equinox occurred April 21st. This is complicated with the Martial equinox of April 30th. On the 18th and 19th a severe marine storm prevailed from Cape Hatteras to Cape May. One ship reports "being surrounded by whirlwinds and waterspouts, culminating in a violent gale from North-northwest, with blinding snow so dense that one could scarcely see three lengths of the vessel." On the 21st, a storm from Western Texas appeared in Louisiana, Mississippi and Alabama; thence southeast over Florida to the Gulf; and thence by the Gulf Stream to New Foundland, where it disappeared on the 25th. When the centre passed by Cape Henry, vessels in Chesapeake Bay experienced the severest hurricane of the Spring and Winter.

The fourth occurred on June 4th. On the 1st a storm centre appeared from the Southwest at Omaha, Nebraska; it moved with very heavy rains slowly to Duluth, on the western extremity of Lake Superior, and was central on the Lake during the 2d; thence it passed northeastward beyond observation. It occasioned severe hail storms, and local wind storms and tornadoes, in Kansas, Iowa, Indiana and Ohio. On the 4th, another low barometer, coming from the Southwest, appeared in Kansas, accompanied by occasional heavy rain. On the 5th it had reached Indiana. Near Winchester, Kentucky, it occasioned a severe hailstorm on the 6th. The ship *Hamilton* reports a severe earthquake at sea on the 4th, Lat. $19°$, $16'$ N., and Long. $57°$, $5'$ W., lasting about ten minutes. During the time there was a tremendous sea on the vessel, pitching her bows under.

## MARTIAL EQUINOXES.

We have collected a large number of facts to demonstrate the nature of the influence exerted by Mars at his equinoxes, and to show the character of the phenomena he may be expected to produce; but we have concluded not to avail ourselves of them at present, for the following reasons: (1) Mars is a small planet, hence his influence must be at best feeble, even if he were not disadvantaged by being placed so remote from the Sun. (2) Since he is a slowly moving planet, requiring only 43 days less than two years to perform one revolution around the Sun, therefore he brings so seldom—only once a year—his equinoctial influence to bear upon meteorological phenomena. (3) His influence—not strongly marked under the most favorable conditions—extends over five or six months; hence it becomes obscured and is easily lost sight of in the frequent perturbations of the planets circulating between him and the Sun; and (4) for the still more cogent reason, that our purpose and labor is more directed to demonstrate by facts that Meteorological Cycles are realities, and what are the causes of them, than to show the nature and energy of the influence exerted by each planet at the critical point on its orbit, and the character of the phenomena we may expect them to produce.

Let the reality of Meteorological Cycles be once accepted as a truth, and then we can, at our leisure, observe, examine, consider and study their phenomena. Then, too, we will learn how to observe and what to observe.

Successful scientific observation depends upon knowing what we want to find.* There is now-a-days much worthless, because aimless, observation. The unscientific man observes and records what attracts his attention in physical phenomena, and as they appear to his untrained observing faculties; but the characteristic facts revealing the nature, cause and laws of the phenomena, being less imposing, escape his attention. The scientist observes minutely the character and varying energy of phenomena, yet not regarding each individual fact as a letter of Nature's alphabet by which she spells out her secrets to Man; he buries the facts beyond resurrection, deep and forever in the grave of averages; a grave that holds relentlessly, like all graves, whatever has been deposited in it; though the progress of Light and Knowledge would rehabilitate it with life and crown it with immortal youth and unfading beauty.

In the Meteorology of the books, we have no established cardinal points whence to start on our voyage of discovery, or whither to push our explorations into the mysterious region of the Unknown, because we have no fixed principle serving as a polar star to direct our course on the wide sea of Investigation on which we embark, and whereon we will be tempest-tossed without compass or rudder, drifting with every tide, or driven

---

Note.—Plato puts it in this shape: "How can we expect to find unless we know what we are looking for?"

by every wind. Hence our meteorological observations, not being animated by a knowing and living spirit, are aimless and lifeless; and are not so directed that they can or will ascertain and fix any great principle in Physical Science. They are all very well, if their purpose be to prove or disprove the phantasy that the supply of fuel to the Sun is varying both in quality and quantity, and that his imaginary fires are dying out. In fact, so far, meteorological observations have only tended to show,—what no really sane man has even doubted,—that the energies of Nature are physical constants, though like the heavenly bodies to whose movements they are intimately related, they are subject to secular inequalities in their manifestations.

It is incontestable that the first step in undertaking the investigation of any subject whatever, is to ascertain and establish one of the fundamental principles of that subject, as a basis of operations to be surveyed and studied, to ascertain what it suggests, and whatever that may be, to follow it upon all occasions. Without an established elementary principle to suggest new and unknown facts, and directing observations for ascertaining them, no progress in scientific research, investigation and discovery, is possible. When progress depends, as it entirely does in Meteorology, upon phenomenal observations, fundamental principles already established must suggest the points to which observation must be directed, because every living principle adumbrates and discloses its next of kin still concealed under a thin veil of mystery.

Now, for a fundamental principle in Meteorology—and consequently as a guide to point out, direct and aid in the observation of meteorological phenomena—we offer planetary equinoxes, which are permanently fixed astronomical events that have occurred at their allotted periods as long as the Solar System has existed, and will recur as long as it endures. The periods—or as the equinoxes mark them—the half periods, of planets, we present as Meteorological Cycles, at the termination or beginning of which, there will be regular recurring phenomena, similar in character, though varying in energy, according to the strength of extraneous causes known to influence the Earth and its Atmosphere at the time.

The great Truth we are inculcating is the reality of Meteorological Cycles, and we present it as a square issue to stand or fall by the crucial test of facts. Hence the more sharply defined and the more distinctly marked the phenomena are, the better they are adapted to our purpose. However interesting as are the Martial phenomena, they are not so well adapted to prove the general principles of Meteorological Cycles, as are those of planets of shorter periods.

So far as my verification of the theory of planetary Meteorological Cycles has been presented, it has been in the order in which it was developed in my own mind. That of Jupiter came first, which suggested and led to the discovery of the Saturnian. This gave me a new view of the Telluric equinoxes; and instead of considering them only as astronomical events, I looked upon them in a new light, and recognized them as marking true Meteorological Cycles. To my mind it was now incontestably established that three of the planets by their passage through their equinoctial points, produced physical disturbances. Might not other planets do the same?

It was ascertained that the Saturnian and Jovial equinoxes produce sunspots, "earth currents," and earthquakes, that is Solar and Telluric perturbations, accompanied with intensely brilliant auroras, which may be considered atmospheric disturbances. The equinoxes of the Earth, besides producing and intensifying the Saturnian and Jovial phenomena, supplement them by adding violent atmospheric perturbations, especially tropical cyclones. When this much was ascertained, all atmospheric disturbances loomed up as phenomena probably produced by planetary equinoxes. Acting upon this suggestion we ascertained the equinoctial points of Venus, and watched the phenomena occurring while that planet was passing those points on its orbit. The result was astounding, and far beyond what was anticipated, fully verifying the suggestion, and moreover unmistakably indicating that bottom had not been touched as yet by the soundings made in that direction. Immediately similar observations were commenced on the phenomena occurring at Mercury's equinoxes, and with equally satisfactory and astonishing results.

But the Mercurial phenomena did not exhaust the Catalogue; for it was soon discovered that not only many phenomena occurred to which it was impossible to assign a place in any of the known cycles, but which, under certain conditions, were so terrifically energetic that they must have a sufficient though unknown cause. All such phenomena were noted, and the first suggestion that they must be referable to Vulcan, the gigantic planet nearest the Sun, soon became a firm conviction. Of this we shall speak presently. This must suffice as an outline of my mental history during the years I sat as an humble pupil at the feet of Nature, learning her alphabet, whilst the subject of planetary Meteorological Cycles was gradually developing, expanding and unfolding itself until I was enabled to grasp it in its fullness.

It was not until the Autumn of 1874 that I turned attention to the phenomena of Mars: consequently I have observed only those of one equinox, that occurring April 30th, 1875; and hence cannot speak so positively as to the character of his phenomena.

Applying the principle to Mars, that a planet's disturbing influence is felt at each of his equinoxes for a term equal to one-fourth of his period of revolution around the Sun, then the disturbing influence of Mars extends nearly through six months. Of course his influence, though quite sensible, well defined, and marked at and near his equinox, will scarcely be appreciable *per se* near the commencement or close of his disturbing period. However it may be with our perceptive faculties, there can be no doubt he does his part, and has his share in the cotemporaneous phenomena.

The influence exerted by a planet is entirely electric. Now, there are two-fold electric states, the static and the dynamic; and the influence of the planet must be so exerted as to produce one or the other, and each alternately upon the Earth. When the static condition is produced on the surface of the Earth, there results from it the dynamic condition on the Atmosphere. When the planetary influence is to produce a static condition on the Atmosphere, there consequently results a dynamic state upon the surface of the Earth. A static condition on the surface of the Earth, induces a dynamic on the Atmosphere; consequently there ensues a down-pour of air from the surface of the aerial ocean; that is a high barometer with *anticyclonal* winds flowing out from all sides of it on the surface of the Earth. A static condition in the Atmosphere induces a dynamic state on the surface of the Earth; consequently an upheaval of air, that is a low barometer with *cyclonal* winds pouring into it on all sides. In Part I this point is fully discussed, and where it is also shown that in Winter a static condition prevails over continents, and in Summer a dynamic. Hence in Winter high barometers over continents are the rule, and low ones the exception; while in Summer the reverse is the case; low barometers the rule, and high ones the exception. Now high and low barometers, as demonstrated in Part I, are electric phenomena; hence a planet's influence being electric, it infuses unwonted vigor during its prevalence into both, thus producing unusually high and unusually low barometers, not only in their proper seasons, but at all seasons, and frequent oscillations and rapid fluctuations in each. Hence in Winter we have those abnormally cold down-pours of air out of a high barometer that smite with death both animal and vegetable life; while at all seasons, but especially in Summer, we have under a low barometer intense upheavals of air and vapor, producing terrific Cyclones, such as hurricanes, tornadoes, waterspouts, etc.

It is yet too early to give the general principles of the modes by which the Earth is affected at a planetary equinox; the general indications, however are, that in the later as well as in the earlier stages, the equinoctial perturbation of a planet affects the Earth so as to produce the static condition, while at or near the equinox it produces the dynamic, mediately as we have just stated by means of the static on the Atmosphere. These effects are much more distinctly marked in Winter than in Summer, or at any other season. In Winter, a week or ten days before the occurrence of a planetary equinox, a high barometer makes its appearance, unusually energetic and accompanied by a downpour and out-pour of intensely cold air, continuing down to within a few days of the occurrence of the equinox. Then there is a fall of the barometer and consequently a rise of temperature, culminating in the sweeping of a storm centre across the continent on the day, or a day or so before, the equinox. This storm centre, bearing rain, sleet or snow, and accompanied with more or less energetic winds, is succeeded by another high barometer, with intensely cold weather. The general influence of Mars, so far as I have been able to determine it, seems to be, in Winter and Spring, to intensify and prolong the normal static Winter condition on continents. Consequently there is a prevalence of high barometers, with their down-pours of cold air from the surface of the Atmosphere. Hence, during the prevalence of his influence, rainfalls in Winter, and frequently in Spring also, are below the average. During Summer and Autumn he seems to intensify and protract the prevalence of the normal dynamic state prevalent in Summer on continents, and hence the rainfalls then generally show a large excess. I attribute the prolonged cold weather of March, April, and greater part of May, and even a part of June, of the present year (1875), to his influence in protracting the Winter static condition. After the dynamic Summer condition had become well established, by infusing more than ordinary energy into it, he contributed his influence in producing the abnormal quantities of rain that have fallen in June and July, up to this writing.

In mid-winter, while calculating the planetary equinoxes for the current year, the crowding of so many of them into June and the early part of July, plainly showed that there would be a turbulent season at this time with heavy rainfalls. A few weeks later I published what I foresaw would be the consequences of such an extraordinary conjunction  But not having taken into consideration the Martial influence, my anticipations fell far short of the reality. There was also another factor I had not considered, namely, that a Saturnian perturbation has been in progress since the latter part of last year, which will culminate in December, 1877.

The special phenomena of Mars that have occurred near his only equinox on which we have made observation, namely, that of April 30th, 1875, and which were either produced directly by him, or indirectly through the intervention of other planets, are Cyclones and Earthquakes; the latter owing probably to the influence of Saturn. The former have not exhibited any extraordinary degree of violence; but some of the latter have been terrific.

The storm of 21st to 25th of April, generated by the Mercurial equinox of 21st, initiated a series of storms following each other in regular succession, until four or five days after the Vulcanian equinox of May 2d. We have already stated that the storm of 21st of April, in its way to the northeast, became, in the Chesapeake Bay, the severest hurricane of the season. It was followed on the 26th by another storm that was attended along its whole path over Louisiana, Arkansas, Tennessee, North Carolina and Virginia, with heavy rains. On the 28th this was followed by a storm from the Indian Territory to the Great Lakes, thence down the Valley of the St. Lawrence, accompanied by " steep barometric gradients," and consequently very high winds, and with protracted rain and snow falls. The winds in the Ohio Valley assumed a decided cyclonic character; and at Erie, Pennsylvania, many buildings were unroofed. At the close of the month another storm appeared in the Northwest. It was central in Illinois on May 1st. It gave rise to local tornadoes from the Northwest to the Lakes, thence South into the Gulf States and Georgia. On the middle Atlantic coast it produced destructive gales. It was immediately succeeded by another storm from the West, and on the 4th followed by a severe storm which on the 6th produced a remarkable hailstorm in North Carolina.

The earthquakes that occurred in the early part of May in Asia and the East Indies, occurred two days after the Vulcanian equinox of May 2d; and those in the latter part of May, three days after another Vulcanian equinox. The earthquake in the Ohio Valley, the Lake Region, and westward, on June 18th, occurred the day after a Vulcanian equinox. The earthquake that occurred on the 15th of May, at Milton, Massachusetts, ElMonte, California, and which wrought such terrible disaster at Cacuta, and other cities in South America, must be considered as a Martial phenomena, owing in degree to the strength that the Saturnian perturbation has already acquired.

Every planet has yet wonderful revelations to make, and when all has been revealed, those of Mars will not be the least marvelous in the catalogue.

## THE VULCANIAN CYCLE.

I have now come to the only point that is vulnerable, if any point is, in the Theory of Planetary Meteorological Cycles; because it is the only point that is not based upon ascertained facts. It takes it for granted that there is an interior planet between Mercury and the Sun, whose position has enabled him to conceal himself so well that with many his existence is yet problematical. If the existence of Vulcan be admitted, then the deductions we have made, taken in connection with the facts that we will adduce to demonstrate and verify them, will be irrefragable.

Astronomers are well acquainted with the fact that Mercury, in his orbital movements, exhibits perturbations that cannot be accounted for upon any other hypothesis than the existence of an interior planet between him and the Sun, just as the orbital motion of Uranus betrayed the existence of an exterior planet, and led to the discovery of Neptune. Astronomers for a long time have been observing and studying the orbital perturbations of Mercury, but as he affords only so few and such short opportunities for observation, not much progress has been made towards achieving their object, which is to ascertain the exact position of this interior planet, that it may be observed at favorable opportunities, and its elements determined. They supposed by this means they could attain sufficient knowledge at least to enable them approximately to determine its transit across the Sun; for that seems to be their only reliance now for attaining a more certain knowledge of its elements. Three or four years ago they thought they had sufficient data to justify them in expecting a transit at or near the Autumnal equinox. They made observations accordingly, but were disappointed.

It was discovered by M. Lescarbault, a physician, on March 26th, 1859. I had the good fortune of seeing it make a transit a few days after the Autumnal equinox of that year, but did not then know of Lescarbault's discovery. Without reflecting upon it at the moment I supposed it to be Mercury, and it was several weeks after before it occurred to me that a transit of Mercury at that season was an impossibility. I then made a record as near as I could, of the date, and its apparent size I recorded as $2\frac{1}{4}$ inches in diameter. As this would give it an enormous size, I have since got those who also saw it, to give me their impressions as to its size, I find they vary from mine a quarter of an inch, some placing it that much more, and others that much less. Probably all these apparent sizes are too great, but they nevertheless clearly establish one thing, that the planet is of gigantic dimensions, and hence from its size and the "energy of its position," so near the Sun, the unavoidable inference is, that it must exert a powerful influence at its equinoxes upon the Earth and the Atmosphere.

There is no reasonable doubt as to the existence of this planet, but nothing is known as to the length of its periodicity. On this point there are mere guesses running from 18 days up to 55 days. As far as I had means of investigating, I could find no foundation for these guesses. I was hence compelled to fall back upon my own resources for determining this point, which I had every confidence could be done at least approximately by consulting facts. With this view I made a classification of all individual phenomena that had occurred in five years, into Jovial, Martial, Telluric, Venusian, Mercurial, and Unknown. The Unknown class consisted chiefly of Cyclones, but there were many auroras, some earthquakes and sunspots. I now ascertained the number of days between the occurrence of the first phenomenon in the Unknown class and that of the second; between the first and the third; the second and the third; etc., and then reduced these numbers to their prime factors. Many of them came out 23, a prime number; and, strangely too, 23 was the greatest common divisor of a majority of the numbers, and those that were not exactly divisible by it were only a day or two in excess, or that many deficient. By comparison, I also found that if 23 be taken as a phenomenal cycle, it cut in at the point and carved out nine-tenths of the most violent and terrific phenomena in the Jovial, Telluric, and Venusian classes. Hence I assumed that 23 days was either the length of Vulcan's period, or half his period. The truth of this deduction—or if preferred, assumption—we will now proceed to demonstrate and verify.

In the month of December, 1784, and January, 1785, there occurred in Iceland the most tremendous volcanic eruption ever recorded in history. "It commenced on the 19th of December, and was accompanied by violent wind and rain, and darkness in the heavens. The whole island shook terribly, and it was feared it would fall to pieces. Three fire-spouts broke out on Mount Skapta, which, after rising to a considerable height in the air, formed a torrent of red-hot lava, that flowed for six weeks, and ran a distance of 60 miles to the sea, in a broken breadth of nearly twelve miles; 12 rivers were dried up; 21 villages totally overwhelmed by fire or water, and 34 others materially injured."

Jupiter's equinox had occurred a little over two years before, and therefore exercised no influence in the production of this terrible physical convulsion. Saturn also was out of the way; and unless Uranus or Neptune (of which I have made no examination) lent their influence, it was entirely the production of Venus, Mercury and Vulcan. The equinox of Venus occurred on the 17th of January, 1785; that of Mercury on the 27th of December, 1784; and that of Vulcan on the 21st of December, or two days after the eruption commenced. On the 13th of January, 1785, Vulcan, in course, added another equinox. It is plain that he not only was the exciting cause, but that he infused new energy into it four days before the occurrence of the Venusian equinox.

In Part I we have demonstrated that Nature, to provide for the wants of vegetable life, has contrived the admirable arrangement of a low barometer during Summer in the centres of continents, to draw and suck into its upheaving vortical column, moist air from the surrounding oceans, to form clouds and rain. Whilst speaking of Mars, we have just stated that, during the prevalence of either a static or dynamic condition, the influence of the planet is to intensify it. It consequently follows that a continent then acts more energetically, under one state it draws down more copious supplies of air from above, and under that of the other, heaves up larger volumes of it than under ordinary circumstances. Hence, when the state is dynamic—as it always is in Summer, and sometimes in Winter—more moist air than ordinarily is drawn from the surrounding oceans into the heart of the continent; and consequently heavier rainfalls than usual are to be expected. Vulcan's size and position near the Sun, give him terrific energy which he never fails to display at his allotted time. One of the consequences naturally following the powerful influences he exerts at one of his equinoxes, must therefore be copious rains; a sequence that has most lamentably been illustrated at his equinox in June, and as we are writing, at that of July, 1875, by the disastrous floods that have occurred in all parts of the Globe. Vulcan, gigantic in size, and fearfully strong in position, with physical laws operating such as our theory postulates, must at his equinoxes produce just such phenomena as the world has witnessed within the last three weeks. Acting singly, his phenomena are of a mild type, but when other planets are accessory, the phenomena are fearful and terrific. The following will show his influence upon rainfalls:

On the 6th of September, 1801, M. Flanguerges obtained at Viviers $14\frac{1}{2}$ English inches of rainwater in 18 hours. Venus passed her equinox August 29th, 1801. Vulcan and Mercury both on September 2d, 1801.

The following fact we present, though not exactly in point, because we want it on record, as it may lead to the discovery of a new point, in the enlargement and perfection of the theory; we want it upon record for another reason: for on several occasions we have found similar facts. The only interpretation I can at present put on these facts is, that Vulcan's periodicity may be 23 days, instead of 46; consequently there will be only an interval of $11\frac{1}{2}$ days between his equinoxes. The characteristics of this class of facts is that they fall midway between Vulcan's equinoxes upon the assumption that his periodicity is 46 days. I have sometimes thought that, like Jupiter, he may produce a major and a minor disturbance, and that so far I have only recognized his major disturbance. Besides the possibility of involving a new principle, the fact bears strong evidence of the correctness of the general theory. It was first brought to the attention of the British Association in 1840, by Prof. Forbes: its authenticity being questioned, at a subsequent meeting, he brought proof of the correctness of his former statement, which was, that on October 24th and 25th, 1822, at Genoa, 30 inches of water fell in 24 hours. The planetary equinoxes occurring near that time were Vulcan, October 13th and November 5th, consequently the rainfall was exactly intermediate. Mercury October 18th, and Venus November 20th.

At Geneva, May 20th, 1827, there were six inches of rainfall in three hours. It being 40 days prior to the occurrence of the next Venusian equinox, we consider her influence null. The equinox of Mercury occurred May 17th, and that of Vulcan May 19th, 1827, the phenomenon was Mercurial and Vulcanian, following the latter's equinox within 24 hours. At Perth four-fifths of an inch of rain fell in half an hour, on August 3d, 1829. Vulcanian equinox August 2d; Venusian August 25th; Mercurial August 31st, 1829. In England, "at Market Laverton, &c.,

hail six and seven feet deep" (drifted we suppose) "fell on September 2d, 1862, doing immense damage to crops." Vulcan's equinox occurred on previous day, namely, September 1st. A Mercurial equinox occurred five days previously, namely, August 29th.

Commander Hall, of the British Navy, thus describes a typhoon that occurred at Hong Kong on July 21st and 22d, 1841: "For days previously large black masses of clouds appeared to settle on the hills on either side; the atmosphere was extremely sultry and oppressive, and the most vivid lightning shot incessantly along the dense threatening clouds, and looked more brilliant, because the phenomena were most remarkable at night; while during the day, the threatening appearances were moderated considerably, and sometimes almost entirely disappeared. The vibrations of the mercury in the barometer were constant and rapid, and though it occasionally rose, still the improvement was only temporary; a storm was therefore confidently predicted. Between seven and eight o'clock in the morning the wind was blowing very hard from the northward, or directly upon the shores of Hong Kong, and continued to increase in heavy squalls hour after hour. Ships were already beginning to drive, and the work of destruction had commenced on every side; the Chinese junks and boats were blown about in all directions, and one of them was seen to founder with all hands on board. The fine basin of Hong Kong was gradually covered with scattered wrecks of the war of elements; planks, spars, broken boats, and human beings clinging hopelessly for succor to every treacherous log, were tossed about on every side; the wind howled and tore everything away before it, literally sweeping the face of the waters. From half-past ten to half-past two the hurricane was at its highest, the barometer at this time having descended to 28.50. The air was filled with spray and salt, so that it was impossible to see anything that was not close at hand; the wind roared and howled fearfully, so that it was impossible to hear a word that was said. Ships were now drifting foul of each other in all directions, masts were being cut away, and from the strength of the wind forcing the sea high upon the shore, several ships were driven high and dry. The Chinese were all distracted, imploring their gods in vain for help; such an awful scene of destruction and ruin is rarely witnessed, and almost every one was so busy in thinking of his own safety, as to be unable to render assistance to any one else. Hundreds of Chinese were drowned, and occasionally a whole family, children and all, floated past the ships, clinging in apparent apathy (perhaps under the influence of opium) to the last remnants of their shattered boats, which soon tumbled to pieces and left them to their fate. On the 26th another typhoon occurred, but not so severe as the first."

As this cyclone occurred 34 days before the next Venusian equinox, the latter must have exercised little if any effect upon it. The same may be said of the Mercurial equinox, which occurred on August 2d. It hence cannot be considered as influenced by either. A Martial equinox however occurred about the middle of June, and still must have exercised considerable influence. Vulcan's equinox took place July 23d, 1841, or two days after the commencement of the typhoon.

A continued gale prevailed at Charleston, South Carolina, with heavy rain from the Northeast, from Monday, 9th of September, 1811, to 10 A. M. on Tuesday, 10th, when it suddenly changed to Southeast. At noon a violent tornado struck the city, passing from Southeast to Northwest in a line 100 yards wide. The centre of the tornado was funnel-shape, and exhibited a lurid appearance, shifting its position rapidly. Many buildings were destroyed and many lives lost; destruction of property immense. It was a spur of a tropical hurricane then prevailing further South. This tornado occurred about eight months before the Jovial equinox of 1812, and twelve days before the Autumnal equinox. Vulcan's equinox occurred on September 9th, the day of the commencement of the storm.

HURRICANE OF OCTOBER 2D TO 9TH, 1842.—We quote this storm for the double purpose of verifying our theory and illustrating points we have several times stated heretofore, namely, the invariable vicinage of a high barometer to a storm centre, or low barometer; and that high and low barometers mutually repel each other. This was a pure Telluric and Vulcanian phenomenon; Vulcan's equinox occurring October 3d, the day after the hurricane was first observed, which was at Tampico, Mexico, on October 2d, where it seems to have originated. On the 4th it was off Balize; on the 5th it covered a large part of peninsular Florida, central a short distance north of Tampa Bay; 6th, at St. Augustine and Charleston; and on the 9th at Bermuda. Immense numbers of sea and land birds were killed and were found floating in the sea. Mr. L. Blodgett,—whose account of it we follow,—states these three additional and important facts, namely, the hurricane went in a more *easterly* direction than usual; that there was a *very high barometer*, 30.10 to 30.46, from the 4th to the 7th, at New York; and that the progress of the storm was *less than ten miles* per hour. These facts prove three points I have repeatedly presented, namely, (1) the invariable vicinage of a high to a low barometer to feed it, that is, to supply it with the immense amount of air it is ejecting through its cyclonal column towards the sky; (2) that when either a high or a low barometer is interposed in the usual route of its opposite barometer, the latter is repelled, the effect of which is either to drive it back, or to swing it around the centre of the interposed barometer; and (3) when such a barometer is interposed the movement of the barometer whose route is foreclosed, is extremely slow.

The usual route of a storm from the Gulf, is northeast, either by the Atlantic coast or the Gulf Stream. Both these routes were closed by the New York high barometer, hence the low barometer or storm centre was deflected out of its usual course, more eastwardly to the Bermudas. Its progress was extremely slow, because there is over the Sargasso Sea, southeast of the Bermudas, a permanent high barometer, between which and the New York high barometer, the storm centre was necessitated to force a passage northeastwardly.

Berlandier describes a hurricane that occurred at the mouth of the Rio Grande, on the 4th of August, 1844, as the most terrible and destructive of any upon that coast. Not a vestige of a single house remained at Brazos Santiago, or at the mouth of the river. The waters of the sea were forced up three leagues from the beach. Vulcan's equinox occurred on the 1st of August, and Mercury's on the 4th of August—the day the storm struck the coast. As this storm came from the Caribbean Sea, it must have existed several days before it struck the coast of Texas, and is a joint phenomenon of Vulcan and Mercury.

One of the most terrible and destructive tornadoes that has occurred in the United States, occurred at Natchez, Mississippi. on the 7th of May, 1840. It was a nucleus in a somewhat general rain. The day began warm and cloudy, with wind south, veering to east. At 2.15 P. M., the sky became a lurid yellow; the storm striking the river six or seven miles below the city, did not reach it until 2 P. M. The rush of wind did not last five minutes, and the destructive blast only a few seconds. Houses were burst outward; 317 persons were killed in the city and on the river. Sheet tin was carried 20 miles, and windows 30 miles.—*Tooley*. The following equinoxes occurred about that time: Venus, May 30th, or 23 days after; Mercury, May 29th; and Vulcan, May 8th, or the day after the tornado. The Medford and Cambridge, Massachusetts, storm, described by Brooks and Eustis, occurred on the 22d of August, 1851, the cloud exhibited the form usually of an inverted cone, though at other times that of an hour-glass, that is, two cones joined at their apexes. The conical point let down from the cloud, moved rapidly about at short distances, now pushing down to the Earth, and now rising from it. Its side motions were compared to those of an elephant's trunk. This action was like the descending tube in a nearly completed waterspout at sea. Its width was from fifty to seventy rods, and its forward motion nearly fifty miles per hour; its duration not over six seconds. Its destruction was unusual in respect to the crushing of objects in its path; panes of glass were perforated and fused by electric discharges, and other evidences of intense electric action were exhibited. Vulcan's equinox occurred on August 22d, the day of the storm.

On the 15th of December, 1851, two enormous waterspouts, accompanied by a terrific hurricane, swept over Sicily; they were described as follows by those who witnessed them. "Two immense cylindrical columns of water depended from the clouds. about a quarter of a mile apart, their points nearly touching the Earth, and moving with immense velocity." They passed over the island near Marsala. In their progress, houses were unroofed, trees uprooted and carried away; men, women and children, horses, cattle and sheep were raised up, drawn into the vortex, and borne on to destruction. During their passage across the island, rain descended in cataracts, accompanied with hailstones of an enormous size; and even masses of ice. Going over Castellamare, near Stabia, it destroyed half the town, and washed over 200 of the inhabitants into the sea. Altogether, 500 persons lost their lives, and an immense amount of property was destroyed; the country being laid waste for miles. After the storm, many bodies were picked up along its path where they had been carried and dashed down to the Earth, frightfully mangled and mashed.

From 1851 to 1874, are 23 years; now every 23 years the equinoxes of Mercury and those of Vulcan occur on the same days of the months. By reference to our table for 1874, it will be found that Mercury's equinox occurred on the 10th, and Vulcan's on the 15th of December, 1874; hence they did the same in 1851, and the phenomenon occurred on the day of the Vulcanian equinox.

The most destructive cyclone ever known in tropical Asia occurred on the 5th of October, 1864. About 100 ships were lost; and over 60,000 persons perished; 43,000 in Calcutta alone. It was accompanied by a "bore" on the Hooghly, the water rising 30 feet, which is 10 feet higher than the highest spring tides, whole towns were nearly destroyed. It indicated its approach for several days, and Capt. Watson, of the *Clarence*, seeing the barometer falling, knew a cyclone was approaching, saved his ship by steering out of its range. Vulcan's equinox occurred on October 1st, 1864, and that of Venus September 19th.

Earth currents are a species of electric phenomena frequently observed in Winter in the after part of a storm centre or low barometer, or rather in the front part of the very high barometer following it, which brings intensely cold weather.

The telegraph wires, in the dark, glow in all the colors of the rainbow; at other times the currents pass along in pulsations, accompanied with fantastic flashes. These are found to be a part of the phenomena attending planetary equinoxes; and important ones they are too; for they are the strongest kind of corroborative proof of the theory of the electric character of physical phenomena which we are inculcating. It is to be sincerely regretted that no accurate record of these phenomena is kept, with all their attendant circumstances. They are not a single letter in the alphabet by which Nature spells out her secrets to

Man, but they are a full formed sentence that she vouchsafes to him for study and analysis.

The following are two instances in which the phenomenon was connected with a Vulcanian equinox. The facts of the first were collected by Mr. W. S. Gilman. We extract only so much as is appropos, from the Annual Report of the Smithsonian Institute, 1867.*

"One of the most beautiful electric phenomena imaginable was witnessed on the evening of the 9th of January, 1868, in the office of the Atlantic and Pacific Telegraph Line, Rochester, N. Y. Suddenly it was discovered that neither wire would work. A continuous current of Electricity was then observed to be passing over the wires and through the several instruments, and this while the batteries were detached. The current seemed to be of the volume of a medium sized pipe stem, and exhibited the several colors of the rainbow. With the key open, the current flowed in waves or undulations; and from the surcharged wire, it leaped over the insulated portions of the key and passed along the wires beyond. The gas of the office was lighted without difficulty by holding the end of the wire within an inch or two of the gas burner. The current was intense enough to shock one holding the wires, or instruments; indeed one of the employees of the office had his fingers scorched by the current. With closed keys the current was continuous."

Mr. Gilman continues: From B. F. Blackall, Manager of the Company at Buffalo, I learn as follows: "At 4.30 P. M., trouble commenced—afterwards located between Fulton and Syracuse—while I was transmitting a telegram to New York over the No. 1 wire; one wire being broken, rested on the ground, and the western end hung across No. 2. At the same instant I noticed my relay surcharged with an unusual amount of Magnetism. Upon opening my key,—which we usually give the sixteenth of an inch play,—discharges of Electricity, averaging as high as 300 pulsations a minute from one platina point to the other took place; and the nearer I placed these points, the more rapid they occurred. The current was passing from West to East through the key. In addition there was a current about the size of a pin flowing from the core of the helices to the soft piece

---

*We are not responsible for the anachronism.

of iron on the armature, which sounded very much like Electricity produced by friction on a glass cylinder, when passing to a Leyden jar. The writer informs me that he has witnessed a half dozen similar displays during the past 14 years." He continues: "From C. W. Dean, Manager of the same line, at Cleveland, Ohio, I learn as follows: An extraneous current made it impossible to work the wire on January 9th last. It was first noticed at 9 A. M., when the current grew so strong that No. 1 wire was opened to Painesville, 30 miles east. This did not help it in the least. I judged that our wires were crossed with those of the Western Union, and that we were getting the full strength of their 100 cups of battery. One thing very strange was that the current pulsated, and the armature of the magnet disconnected from the battery and the wire open east, vibrated like a pendulum."

He furthermore says: "From J. A. Osborne, Buffalo, New York, I learn that the wires of their office were so heavily charged that he thought certainly they were crossed by the Western Union wires. They could not be touched. The current passed over in waves; and it was necessary to throw the instruments out of circuit to prevent damage to them. Fantastic streaks flashed across the wires. At one time a continuous stream of fire passed off which lasted for four or five seconds. At Lockport the Electricity set fire to a board to which the wires were attached. The magnets became so surcharged with Electricity that when the wires were disconnected the armature remained drawn up to the coils for full three quarters of an hour." Mr. Gilman gives the following as the state of the weather at the time: "Sky clouded at Rochester, Toronto and Montreal, and storming."

We have demonstrated in Part I that high and low barometers are electric phenomena, and that a high barometer is a descending, and low barometer an ascending current of Electricity. We also show there that the phenomenon of "earth currents" is Electricity flowing from an area of high to an area of low barometer; and above we have stated that the phenomenon makes its appearance in the after part of a low, or the fore part of a high barometer, which on account of its more than ordinary energy brings intensely cold weather with it. Prof. Henry's note to Mr. Gilman's paper says: "On the night in question an aurora is noted at Independence, Iowa, and a heavy snow in Michigan. A wave of low temperature was passing from West to East, from the 7th to the 10th of January, reaching its minimum in the State of New York on the 9th and morning of the 10th." Now any one who has watched and compared the readings of the mercurial movements in the barometer and thermometer, knows that to say that "a wave of low temperature was moving from West to East," or in any direction, is the same as saying that a high barometer was moving in that direction. Our proposition therefore is verified by the fact stated by Prof. Henry. By referring to the table of Planetary Equinoxes at the end of this volume it will be seen that Vulcan's equinox occurred on January 9th, 1867, the day of the phenomenon.

According to newspaper accounts, on the afternoon of January 8th, 1875, intense electric currents filled the telegraphic wires from Wyoming, eastward to Iowa; streams of fire seeming to flow along them; and when open, the keys were either surrounded by a halo of light, or discharging so rapidly as to be an almost unbroken stream of sparks. In the centre of the Mississippi Valley, and North, a general snowstorm prevailed, and rain in the Southern States. At St. Louis, about noon, it turned intensely cold, with a fierce and piercing gale from the northwest, the front of an approaching very high barometer, ushering in an intensely cold spell of weather. By reference to our table of equinoxes, it will be seen that Vulcan's equinox occurred on the day previous, January 7th. The Weather Review of the Signal Service shows the following to have been the condition of the weather and cotemporaneous phenomena: On the 6th, a low barometer announced itself in the South Atlantic and Gulf States by rainy weather and northeasterly winds. Thunderstorms prevailed in Georgia. It was not until the 7th that its movement could be definitely traced; and on its way northeastward, along the coast, it became a dangerous rain and snowstorm. On the morning of the 8th, the barometer at Halifax, Nova Scotia, stood 29.4 inches. Rains, sleet and snow, principally the latter, fell from Tennessee, Kentucky, Indiana, and Lake Huron, eastward over the Lower Lake Region, to the Middle States, thence northeast to the Gulf of St. Lawrence.

Storm centre, or low barometer, No. III, developed in Oregon on the 6th. On the 7th it appeared in the Upper Mississippi, and was central at St. Paul on the morning of the 8th. Its advance winds, those blowing into it from the ocean, became dangerous on the coast of Maine, and northward. It was closely followed by high barometer No. 3, with high and dangerous winds blowing into it—the low barometer—from the west. In Texas this caused a severe " Norther ;" velocity at Indianola 52 miles per hour; on the Lakes, 40 to 45 miles ; at Cape May, 40; Sandy Hook, 67 ; and Eastport, 42 miles per hour. Snow accompanied it in the northern sections, and light rains in the Southern States. High barometer No. 3 made its appearance in northwestern Manitoba, on the 7th. On the night of the 8th it was at North Platte, in the extreme western part of Nebraska. It will thus be seen that the electric current in wires from the West, in the afternoon of the 8th, was, as we maintain, a transmission of Electricity by an earth current from the advancing high to the retreating low barometer.

As it is now over three years since in writing the chapter treating on this subject in the *"Elements of Meteorology"* we first postulated this theory of electric earth-currents, we may as well, since the matter is up and a new fact at hand, produce it in verification of the theory. In the Monthly Weather Review, of January, 1875, it is stated that the observer at Santa Fe, New Mexico, reports, that on the 15th of January, an extraordinary electric storm on the Telegraph was noticed from 12 M. to 3 P. M. The current was so strong that the line could not be worked ; the key was left open and most of the time was surrounded by a ring of fire. The Review states that this was during the passage of low barometer No. VII* that this happened. Storm centre No. VII, in consequence of high barometer No. 5 covering Ohio and the region eastward, was deflected, and took the abnormal direction from the Northwest to the Southeast, and in some part of its course even west of South. At noon of the 15th it was nearly due east of Santa Fe, and the latter locality was falling under the influence of high barometer No. 6, then

---

*NOTE —The Review says No. VIII, which evidently is a typographical error, since that low barometer did not appear until the 19th, four days after, and then in Oregon.

appearing in Wyoming. That this was the case I adduce the facts stated by Prof. Loomis before the National Academy of Science at Washington. His statement in substance is as follows; "At Denver, January 14th, 1875, the thermometer, with a variable northeast wind, had been below zero all day." Before quoting further we will remark the cause of this cold northeast wind at Denver was high barometer No. 5, then covering Nebraska, consequently sending out flows of cold air in all directions from its centre. Professor Loomis continues: "At 9 P. M. the thermometer was one degree above zero. The wind then shifted suddenly to the Southwest at 9.15, and the thermometer stood at 20°; at 9.20 P. M. at 27°; at 9.30 at 36°; at 9.35 at 40°; after which there was little change, till near noon next day, January 15th, when with a fresh Southwest wind it rose till 11.30 A. M., when it stood at 52°. *The wind then suddenly backed to Northeast and at 12.30 P. M. the thermometer* had fallen to 4° *above zero.*"

Now, what is the explanation of all these extraordinary facts? A few moments examination of the relative positions, at that time, of the synchronous high and low barometers, as shown upon Map I, of Monthly Weather Review for January, will make it so plain that, "he who runs may read." High barometer No. 5, that had been pouring out its cold air over Colorado, had, at the hour of the change of wind, passed over the Missouri River, south of Omaha, into southern Iowa. Low barometer No. VII, advancing from the northwest, was now in Wyoming. An inpour of air into it was necessitated from the nearest high barometer, probably situated in southern Arizona, or it may be, in Mexico; this brought about a sudden change of temperature. This warm outpour of air continued to flow until low barometer No. VII had passed. But it is closely followed by high barometer No. 6, from the northwest, now already in Wyoming. Consequently, a sudden swing of the wind takes place from the warm outpour of a southern high barometer, to that of the gelid outpour from a northern high one coming up from the Arctic Circle and drawing its supply of air thence.

Now, mark it, at 12.30 P. M. this cold outpour of the advancing Arctic high barometer had, in less than one hour, lowered the temperature of the Atmosphere at Denver 48°, for at 11.30

A. M. it stood 52°. Now, it is precisely at 12 M. on that day that the electric current in the wires is discovered at Santa Fe.

There can be no doubt that the theory we have advanced concerning these earth currents is the true one; for as far as we had opportunity and *data* for investigation, the facts invariably have verified it. We think this a point of the highest importance in our general theory; hence we could not pass by a fact so pointedly confirming it, without availing ourselves of its testimony. For be it known, that whatever proximate causes, such as high and low barometers, planetary equinoxes, etc., we may assign for physical phenomena, we hold the ultimate, or the cause of causes to be Electricity. Hence we consider it all times in order to adduce a fact when we have one that establishes our fundamental principle, namely, that Electricity is the cause and common bond of union between physical phenomena. Here was a fact that not only proved this, but it did more, it showed that high and low barometers are parts of an inseparable couple or pair, and that Electricity is the bond of union between the parts.

We here repeat in substance what we have said in Part I on this point: that Electricity is a polar force. It developes in pairs; as it cannot exist singly, so it cannot act singly. It therefore follows of necessity wherever there is electric action, the *phenomena* cannot be simple, but must be duple. One part of it must be in one state, and the other part in the opposite state. The negative state always exists on the surface of the Earth, and the positive on the clouds and Atmosphere. But both of these states may be alternately static and dynamic. When the negative state on the Earth is static, it draws down, or rather compels the dynamic on the Atmosphere to pour down upon it an immense column of oppositely electrified air, creating as it were in the æriel ocean an immense maelstrom. This descending column of air causes what is popularly called a high barometer. When the static is on the Atmosphere, it draws up, or rather compels the dynamic on the surface of the Earth to throw up an immense column of negatively electrified air. This produces the phenomenon commonly called a low barometer. Low barometers and high barometers therefore are parts of electric couples in the phenomena not only of atmospheric circulation, but in the distribution of rain and sunshine over the

Globe; and here we find in these Earth currents Electricity, as might be anticipated, flowing from the high to the low barometer.

After this long digression, let us, like the French, say: *"Revenons a nos moutons,"* which is to verify the Vulcanian Cycle.

On the 23d of December, 1854, a severe and destructive earthquake occurred at Jeddo, or Yeddo, the capital of Japan. On the 11th of November, 1855, another, and the most terrible earthquake ever experienced in those Islands occurred. "During it, 57 temples and 100,000 houses were destroyed, and over 30,000 persons lost their lives."—(*Haydn*.) Since, in every 23d year, Vulcan's equinoxes occur on the same day of the month, therefore the sums of $1854 \times 23$ and $1855 \times 23$ show that the Vulcanian equinoxes of the years 1854 and 1855 were on the same days of the months as they will be in the years 1877 and 1878. By inspection of the table of equinoxes in Appendix, it will be seen that an equinox of Vulcan occurred on the 24th of December, 1854, or the day following the first earthquake, and that a Vulcanian equinox occurred on the 11th of November, 1855, or on the very day of the terrible catastrophe.

Mr. Nicholas Pike, United States Consul at Port Louis, gives an account of a hurricane the United States steamer Monocacy encountered in a passage from Simon's Bay, South Africa, to Mauritius, from which we extract the following "The whole of our passage, since leaving Simon's Bay, had been a succession of bad weather, and a few sunny days, which we in reality had, were to both officers and men a veritable blessing; sails were repaired, clothes dried and mended, and the decks for the first time quite dry, resounded with the joyous laughter of the crew. But their joy was of short duration." This was in the early part of January, 1867. If the reader will refer to our table he will find that a Mercurial equinox had occurred December 27th, 1866, and therefore the narrative relates what took place in the interval between the Mercurial equinox and a Vulcanian which occurred on the 6th of January, 1867, in fact on the eve of the latter. Mr. Pike continues: "On the evening of the 6th of January, the sky became gloomy, dark and threatening clouds passed swiftly to the northward, the sea rose fast, and the vessel commenced to roll heavily, bedding and clothes were quickly taken below and everything secured for bad weather again. The night from the 6th to 7th of January, fully justified our anticipations; heavy blasts of wind, rain and lightning, the rolling of the vessel, the cracking of her timbers, and the thundering noise of a wave breaking under the vessel's counter, made, I may safely say, even the oldest seaman on board uncomfortable; especially as the vessel being new, and her seaworthiness to all, even the Captain, unknown, we had not the confidence in her with which her gallant behavior afterwards during the following gale inspired us all; sails were reduced, or partly so, by the aid of the storm, the flapping of the canvass, torn to ribbons by the rage of the tempest, the loud thunder, the occasional flashes of lightning, the rising of a tremendous wave, showing first its white foaming crest far off in the horizon, and then drawing nearer and nearer, till you might almost fancy it would instantly engulf us, but our gallant ship rose nobly to the crest of the surge. All this was a spectacle wild and fearful to behold, but in its very wildness, grand and sublime. The men worked hard, sending down masts and yards, repairing or bending the storm-sails, or standing at the pumps, knee-deep in water that washed unceasingly over the decks. Daylight showed us at last the extent of damage the vessel had sustained. The paddle-boxes, the round-houses were smashed in and washed away; the rail forward was stove in, and the heavy one inch iron plates were bent double; ring-bolts, to which the heavy pivot guns were secured, started from the deck, and the guns threatened with each roll to break adrift from their lashings; a temporary lull in the gale gave us time to secure them, and repair damages a little. Everybody hoped for good weather, as the heavy rain which fell during five hours beat down the sea considerably. But on the evening of the 7th the storm commenced again. A red lurid light spread all over the sky, and shortly after the setting of the Sun, the ocean became furious once more. A tremendous sea, breaking over the starboard bow, swept everything before it, tearing away the gratings of the hatches, breaking the after sky-lights, and rushing down into the wardrobe and cabin, floating and drenching everything and everybody. The tiller ropes having been carried away, paying off before the wind, became unmanageable; the guys of the smokestack having broken, it was feared that the heavy mass of iron would descend upon us, smashing everything; the ship then coming to again filled her decks with water, and leaning over to port remained so long in that position that even the stoutest hearts quailed, and anxiously counted the seconds, till at last the ship rose gallantly again on the crest of the next wave; luckily the sea had stove in the lower ports, the immense quantity of water found a ready egress from her decks; and the vessel, lightened of her weight, rolled less heavily; new wheel-ropes were rove, and the storm having spent its fury abated greatly. A little before six o'clock the Sun rose red and gloriously in the East, in a serene and cloudless sky."

In order to give as many facts in verification of the theory, we will condense them as much as is compatible with intelligibility.

1866—June 13th, Vulcanian equinox; June 15th, hurricane at Winona; 17th, hurricane in New York city.

1867—Vulcanian equinox May 1st; tornado at Tuscaloosa, Alabama, April 29th.

Waterspout in Lake Michigan July 31st. Vulcanian equinox August 1st.

1868—March 17th, hurricane at Chatham, Illinois, and storms generally in the West. Vulcan's equinox March 19th.

May 3d, tornado near Muscatine, Iowa; 4th, a hurricane passed from Mississippi, south of Columbus, eastward through Pickens and Tuscaloosa counties, Alabama. Vulcan's equinox May 4th.

1869—April 18th, severe hailstorm at St. Louis; tornado at Dubuque and Burlington, Iowa, and Indianapolis, etc. Vulcan's equinox April 14th; equinox of Venus April 30th.

May 28th—Tornado at Athens, Ohio; destructive hailstorm at Wheeling, West Virginia. Vulcan's equinox May 31st.

1870—Gales on Lakes January 17th. Vulcan's equinox January 15th.

April 17th—A severe rain and snowstorm from the northwest into the Gulf States; snow at St. Louis six inches deep; thermometer fell to 27°; at Springfield, Missouri, to 18°; at Oxford, Mississippi, to 28°; cotton and corn cut down even with the ground. Vulcan's equinox April 17th. Severe gales and heavy rains in Europe on the 19th of April.

October 15th—Tornado at Milwaukee; terrible tornadoes in Ohio and Kentucky. 21st, tornado at Belleville, Ohio. Vulcan's equinox Oct. 18th.

1871—The phenomena of February to middle of March have already been given. March 28th, Vulcan's equinox. A cyclone at Auckland, New Zealand, on the day of the equinox.

Vulcan's equinox April 20th; on the 18th violent hailstorm at Leavenworth, Kansas, and at St. Joseph, Missouri, accompanied with terrific winds, rain, thunder and lightning; same day a severe hailstorm at Jackson, Mississippi. 19th, a severe hailstorm at St. Louis, and in Ohio; and an earthquake at Oxford, North Carolina; 23d, a general and destructive frost over the country.

Vulcan's equinox May 13th. On May 14th, destructive tornado at Mosinee, Wisconsin, followed by a general rain of nearly a week's duration.

We have heretofore given all the Vulcanian phenomena from June to the end of October, 1871, in the phenomena of the other planets, except the following: At Vulcan's equinox of July 21st, a large sunspot seen for several days, attained its maximum on the 21st. On the 17th, terrible thunder and lightning near Burlingame, Kansas, all vegetation killed for 20 to 30 yards around where it struck. 20th, earthquake at Santiago, Chili, and a general earthquake in the New England States.

At the equinox of Vulcan, Nov. 13th, 1871, the following phenomena appear on my record: A very brilliant aurora from midnight to daylight in the morning of Nov. 10th, and again at night. On the same day intense magnetic disturbance were observed at the Observatory at Havana. On the same day a severe snow storm from Lake St. Clair to Nova Scotia, extending south over Vermont, New Hampshire, and Maine. On the 13th, severe gales in Oregon, very destructive to shipping; general rain and snow storm in the Mississippi Valley, with violent gales on the Lakes and Gulf. November 14th—A violent hurricane in New Jersey, New York and Boston, doing much damage both on sea and on land. In the Signal Office Report of 1872, there is a minute description of the snow storm from the Northwest; we extract the following items: Barometer fell .36 inch at Omaha from the 11th to the 12th; temperature rose 18°, and rainfall .7

inch. Leavenworth barometer on 11th, 30.4 inches, on the 12th, 29.95 inches; temperature rose 25°; rainfall 1.1 inches. St. Louis, on the 12th, barometer 30.38, on the 13th, 29.9 inches; temperature rose 15°; 1.94 inches of rainfall. Chicago, barometer on the 12th, 30.48; on the 13th, 29.66; rain, 1.18 inches. Memphis, barometer on the 12th, 30.31, fell to 29.62; rainfall .5 inch. Similar variations took place in the Ohio Valley: the following was the rainfall at the principal stations: Indianapolis, 1.93 inches; Louisville, 1.16, and Cincinnati, 2.21 inches. At the stations along Lake Erie the barometer fell as follows: At Detroit, from 30.58 to 29.48; at Cleveland, from 30.58 to 29.35; at Buffalo, from 30.62 to 29.23; at Rochester, from 30.63 to 29.25. The rainfall at these stations ranges from 1.54 to 2.70 inches. The paper concludes by saying that this was the severest storm that occurred for many years on the Lakes; and that at the same time this storm was on the Lakes, another storm was moving down the Atlantic coast. The intensity of this storm is not entirely owing to Vulcanian influence, for Jupiter's perturbation was yet at its height, he having passed his equinox only 47 days before.

1872—The first Vulcanian equinox in this year occurred January 21st, 1872. A Mercurial equinox having occurred three days previous, the phenomena at this period are complicated. A very large sunspot first showing itself on the 13th, attained its largest size on the 21st. At many of the physical observatories in both the Northern and Southern Hemisphere, strong earth-currents were observed during this time. Earthquakes were frequent and violent in all parts of the world. On the 14th and 15th there were severe shocks in the Himalayas, which were also felt at Broosa, Asia Minor. At noon on the 15th two severe shocks were felt at Quebec. On the 16th, the city of Shamaka, near the Caspian Sea, was totally destroyed: and a severe shock also on the same day occurred at Valparaiso, Chili. On the 22d there is an earthquake recorded at Guayaquil, Ecuador. The reader will bear in mind that the period at which these earthquakes occurred, was only four months after a Jovial equinox; and they not only illustrate but verify the principle we have often laid down, that it is the equinoxes of the minor planets that call forth the most violent physical paroxysms, during the perturbations of the major planets. The series that occurred, as detailed at this period, were evoked by Mercury's equinox on the 18th, and by Vulcan's on the 21st. The previous equinox of Vulcan, that of December 29th, 1871, brought about that terrible, or rather series of terrible earthquakes in Persia, in which Khabooshan, in the Northwestern Khorassan, was destroyed, and over 30,000 persons are said to have perished.

On the 18th there was a snowstorm in the Mississippi and Ohio Valleys, with heavy rains South. On the 23d a remarkable snowstorm occurred in Colorado. The fore part of the day had been unusually warm and clear. An hour before sunset an arctic snowstorm suddenly burst from the Northwest, covering the ground from 15 to 18 inches. At Greeley it was accompanied by a terrific gale, and intensely cold. Several persons who had gone beyond the Cache La Poudre for coal, perished with their teams, and large numbers of cattle and other stock perished with cold.

The second equinox of Vulcan in 1872 occurred February 13th, and is complicated with that of Venus, which occurred on the 5th; we may therefore as well give the general as the special phenomena as entered on my phenomenal record. January 29th: A hurricane, or a series of hurricanes, was started in the Indian Ocean on this day, which continued to the 16th of February, when last encountered and observed by a ship, but may have continued longer. February 2d, it snowed all day. Serene all day on the 3d; at 4.10 P. M., I observed the beams of an aurora in a perfectly serene and blue sky in the magnetic North, diverging from a point about 15° above the horizon. Streamers appeared like sunbeams behind a cumulus cloud. I predicted a storm within 24 hours. The aurora continued into the night.

On the 4th it snowed all day; an aurora illuminated the clouds all night. This aurora was extremely brilliant where the sky was clear. It was seen in the Eastern States, in England, in Australia, and in New Zealand, followed by heavy snows in America, and rainstorms in Europe and in the Southern Hemisphere. At Manchester, England, intense magnetic disturbances observed from 4 to 11.30 P. M., the deviation amounting sometimes to as much as 6°. At Mauritius, and at observatories in Tasmania and Australia, an intense "magnetic storm" prevailed from 4th to 5th, and terrific hurricanes were encountered by ships on both days on the Indian Ocean. On the 5th, the severest snowstorm of the season at St. Louis; and a general storm over the entire continent. On the 6th, three distinct and severe shocks of an earthquake felt in Michigan, and at Winona, Minnesota; a severe shock was also felt in the Herzegovina. On the 7th, *the Sun was covered with large spots.* (Query: Was this the effect of the Venusian equinox?) An earthquake at Cairo, Illinois, at 4 A. M., on the 8th; and general and heavy rains in Oregon and California. On the 12th it was warm and clear, but unmistakable indications of an approaching storm. Tremendous snowstorm during the night in Minnesota, and extending southeast to Central Illinois. On this day hail fell throughout the province of Kattywar, India; a hitherto unknown phenomena in that country. On the 13th, thunder and lightning during the night in Missouri and southwest, ending during the next day with a furious cold gale from the northwest. The thermometer fell $51°$ in 12 hours. Heavy rains south, and a fearful gale at Rock Island, Illinois. In the forenoon there was an earthquake at Lisbon. On the 14th, very cold, with the thermometer $2°$ below zero at 6 A. M. According to telegrams, it was warm and raining heavily in Utah. On the 15th more moderate at St. Louis; clear, with wind from the southeast, with every indication of an approaching storm. On the 16th it commenced raining at 4 A. M., turned to snow before day, and snowed until late in the evening.

The next Vulcanian equinox in order is that of March 8th. In the lake disaster list we find as follows: March 7th, steamer *Ironsides* detained by ice during a violent gale on Lake Michigan; schooner *Challenge* sprung a leak; propeller *Missouri*, loaded with grain, injured, leaking badly; *Manistee* aground and goods thrown overboard; etc. March 6th, a severe shock of an earthquake was felt at Dresden, and other parts of Germany, between 4 and 5 P. M. 10th, an earthquake destroyed the seaport town Hamade, and the capital of Sekishu, Japan; over 500 lives lost. It is said that a mountain was thrown into the sea, and everything subverted on land.

Another equinox of this planet occurred on the 31st of March. On the 27th a destructive cyclone occurred in Australia. In the lake disaster list we have the items: March 30th the schooner *North Star* stranded in a snowstorm at night; schooner *Two Brothers* driven ashore; schooner *C. L. Johnson* damaged in a a gale; tug *Margaret*, and scow *Rough and Ready* sunk. On the 31st the propeller *St. Joseph* had furniture injured by rough weather. On the 26th the terrible earthquake at Inyo, California, commenced, which continued to April 13th. On the same day an earthquake was felt at Paducah, Kentucky. This earthquake was felt all along the Pacific coast to the city of Mexico, and beyond to Oaxaca, where many churches and houses were levelled to the ground. On the 3d of April the city of Antioch, Syria, was destroyed by an earthquake.

Vulcanian equinox No. 5, of 1872, occurred April 23d. The lake list of disasters again shows damage from 21st to 26th, but we cannot quote them. In the Annual Report of Signal Service for 1872, a storm of 24th of April is mentioned as one of those of which timely notice was given to the shipping interest; but there are no particulars given. On my phenomenal list I find, April 21st, a cyclone in South Carolina; 62 houses destroyed in Chester, and great damage done at Columbia. An earthquake felt at Memphis, Tennessee, and Cairo, Illinois, on the 23d. A gray haze and lurid Sun to-day, indicating the imminence of earthquakes. 24th, gray haze, lurid Sun, and unseasonably high temperature. This remark is appended, "I would feel alarmed if I lived in an earthquake country." 25th, foreign dispatches state that an earthquake occurred at Dumphries, Scotland, on the 23d, and that yesterday (24th) a terrible eruption of Vesuvius took place. A violent series of earthquake shocks commenced on the 25th, and continued to the end of the month, at St. Salvador, Central America. An earthquake at Attok, East Indies.

Vulcan's equinox No. 6 of 1872, occurred May 16th. The lake list again furnishes a list of disasters, and on the 18th it says, at 11 o'clock A. M. a storm at Buffalo drove vessels back to port.

Equinox No. 7 of 1872, occurred June 8th, 1872. This equinox is complicated with that of Venus of May 30th, attendant with the usual phenomena of rainy weather, frequent storms, and in this instance an unusual number of water-spouts. The leaves of my phenomenal record, covering the greater part of the month of May and part of June, have become detached and lost, consequently I am unable to give phenomena until June 16th, where I find, "Weather very showery and numerous heavy local rainfalls. They seem to have been frequently of the nature of water-spouts. At St. Louis the highest cirrus clouds had a decided westward movement. The cumulus clouds were about stationary, or if any movement it was towards the West. They would gather suddenly into a large mass, then change rapidly into nimbus and pour down a heavy local shower until exhausted; after a very short interval, another similar knot would gather to the westward and pass through the same phases. These knots seemed to propagate themselves in streaks from East to West, and after disappearing in the western horizon, similar knots reappeared in the East. These local showers were attended with thunder and lightning, hardly ever more than one discharge before the cloud became exhausted." This remarkable character continued until after the terrific thunderstorm at St. Louis, on the night of the 27th, which we have described in Part I. The phenomenon was so remarkable that I made a minute and full record of it at the time, because I was anxious to know what it meant. I was so impressed with it, that a few days after, while the dispatches were bringing in detailed accounts of extraordinary phenomena from both East and West, I added this memorandum: "Here, at St. Louis, the singular peculiarity was to develope in knots or points from five to eight miles apart, and on a line running from East to West; a knot consisted of a cumulus cloud which rapidly passed into a nimbus, and exhausted itself in a short downpour of rain. Scarcely had it become exhausted before another knot was seen forming westward. There were always two lines of propagation visible, one North and the other South; but at one time when a line passed through the Zenith, it was flanked by a line on each side. The lines seemed to be about 20 miles apart; and they were renewed by pulsations or waves from the East, at intervals of about every three hours." Judging by the phenomena afterwards reported, a similar condition must have prevailed over the whole Continent at the time. In the Eastern States a number of extraordinary local rains with floods were reported as having occurred on the same day. At St. Joseph, Missouri, a terrific rain, hail and wind storm occurred, producing a flood that did immense damage. On the Western Plains there were what are popularly called cloud-bursts. About a month after, in eastern Colorado, I saw the track made by one of these moving columns of water over the Plains. In a line almost straight, it had scooped out a row of holes from three to ten feet wide, and from two to five feet deep. Near Golden, on the same day, a water-spout fell on the mountain, by which a Mr. Virden and his family, returning from church in a carriage, through Clear Creek Canyon, were overwhelmed, and Miss Virden and Miss Blood lost their lives. Their bodies were found several hundred yards below, covered with boulders, gravel and sand, horribly mangled. Mr. and Mrs. Virden saved themselves by grasping a limb of a tree, and holding on until the wave had passed, which was twelve to fifteen feet high and nearly perpendicular. In southeastern California, and Nevada, on the same day, extraordinary sand-spouts were seen, six and eight in number, at the same time, in a row, spinning and whirling like water-spouts at sea. On the Central Pacific Railroad, near Truckee, there was a water-spout, or as the people of the mountains expressively call such a phenomenon, " a cloud-burst." When first seen by the inhabitants, it was a funnel-shaped cloud of a large size, and black as pitch. It was upright in position, and had a swift whirling motion. It at first descended within fifty yards of the ground; it then lifted a little, and swooped by with immense velocity for about half a mile, when it struck and fell upon the railroad, washing it completely away, embankment, superstructure, and all, and in many places twisting the rails. In England, two days after, namely, on the 18th of June, a most terrific rain and hailstorm occurred at Birmingham, accompanied by dreadful thunder and lightning. Large masses of ice fell, and the amount of rain that fell was estimated at 250 tons to the acre.

I state these facts because they have a peculiar interest to me. I recorded them at the time of their occurrence, because I wanted for study and comparison a record of consecutive individual facts, uncontaminated by the touch, or distorted by the whims, caprices and prejudices of Man. I wanted to know whether such extraordinary phenomena as these were normal; whether they had a periodicity; and what was the length of their period. If they had a period of a definite length, then I wished to ascertain the fixed cause that regularly brings them around in their circuit. I love and venerate facts, because they enabled me to penetrate the veil that hid the mysteries of the seasons, and taught me to interpret the hieroglyphics written upon one of the tablets in the Great Temple of Nature.

Vulcan's equinox No. 8, of 1872, took place on July 1st. On this day, Balasore, in India, was nearly destroyed by a cyclone. A storm centre passed over Lake Superior on the 30th of June, succeeded by another that appeared in the northwest on the same day, and was central in the Ohio Valley on July 1st.

Vulcan's 9th equinox of 1872, took place July 24th. The lake disaster list, amongst others, mentions the following casualties: On the 22d, tug *Harrison* capsized; barge *Emmett* lost part of deck load in a squall; a raft of timber broken up by heavy weather. On the 23d, the schooner *Couch* sunk; schooner *Collier* lost main-topmast in a squall.

Vulcan's equinox No. 10 of 1872 took place August 16th. On the 15th, schooner *Dayspring* struck by lightning on Lake Michigan; one man killed, and another seriously injured. 16th, bark *Saginaw*, loaded with timber, sunk. August 15th, a brilliant aurora to-night. One of the wires of the Western Union Telegraph Company was worked some time by the auroral current. On this day a tornado swept from East Longmeadow to Wilbraham, leveling everything in its path for the distance of five miles. Stone walls and fences were strewn in all directions. A strip from five to fifteen rods in width was cut clean through a forest of large trees, and several buildings thrown down.

A re-examination of my data for the purpose of culling the Vulcanian phenomena, discover the following facts, overlooked when I made up the Mercurial record. As they are European, and also too important to be omitted, we add them here, since it is too late to insert them at the proper place. By examining the Table of Planetary Equinoxes, it will be seen that a Mercurial equinox occurred August 26th, 1872. On the night of August 25th a water-spout burst above the coach road near Loch Katrina, Scotland, rendering the road impassable for several days, by the trees, debris, etc., swept down by the flood. The Birmingham (England) *Morning News* says: "The people living near King's

Sutton, Banbury, say that on the 25th inst. (that is August 25th, 1872) about one o'clock, they saw something like a haycock revolving through the air, accompanied by fire and dense smoke. It made a noise resembling that made by a railway train, but very much louder, and travelled with great rapidity. It was sometimes high in the air, and sometimes near the ground. It passed over the estates of Col. North, M. P., Sir W. R. Brown, Bart., and Mr. Leslie Melville Cartwright, whose park wall it threw to the foundation in several places, and at one place for upwards of sixty yards. A man named Adams was breaking stones, and a moment before he was standing under a tree, which was suddenly torn up by the roots and the branches scattered in every direction. Two or three other trees near him were also torn up, and one of them, the largest, a beech, on Sir William Brown's estate, tore up with it 12 or 15 tons of earth. For a distance of nearly two miles, hedges, rails, fences, trees, hovels and ricks have been knocked down or injured. A whirlwind followed the fire-meteor, and carried away everything before it. Stones from the walls knocked down were carried forty yards away, and the water in a pond on the passage of the phenomenon was sucked up. After traveling about two miles the meteor seemed to expend itself, and disappeared all at once. There was a heavy fall of rain at the time and a vivid flash of lightning just before." On the same day there was a destructive tornado in Ireland.

The 11th Vulcanian equinox of 1872 took place on the 8th of September. On the 8th of September, according to Mr. Denning, the disc of the Sun was covered with spots. On the 9th, 7.2 inches of rain fell at Bombay. Heavy rains continued for several weeks in India and Bombay, sometimes accompanied by severe cyclones and destructive floods, by which much property was destroyed and many lives lost. September 9th and 10th, a severe hurricane at St. Kitts, Barbadoes, and other islands. The observations of the Signal Service show for the 8th of September, rain at the following stations: Omaha, Keokuk, Davenport, Detroit and Burlington, Vermont; and heavy rains at Punta Rassa, Chicago, Milwaukee, Oswego, Mount Washington and Quebec.

The 12th Vulcanian equinox of 1872, occurred on October

1st. On the morning of the 28th of September, a low barometer was central between Omaha and Leavenworth; at 4 P. M. it was central at St. Louis, and at midnight near the southeastern point of Lake Michigan. On the morning of the 29th, the centre was in Northern Michigan, northwest of Saginaw, thence it passed northeastwardly over Lake Huron, during the day. On the morning of the 30th it was central over Montreal, thence it passed southeast into the Atlantic on October 1st. It reappeared on the coast of Maine on the 2d, and on the 3d extended into the Valley of the St. Lawrence. This low barometer was attended with heavy rains from Omaha to Quebec, and from Norfolk, Virginia, to Nova Scotia.

Another storm centre developed in the northwest on the 2d of October, and passed over the Lake Region on the 4th, accompanied with light rain from the Rocky Mountains to Florida, the Lake Region and St. Lawrence Valley.

The 13th Vulcanian equinox of 1872, took place on October 24th. The Signal Service Report of 1873, p. 981, says: "The most severe storm was that which from the 21st to the 27th, travelled from the Gulf northeastward, over the South Atlantic and Middle States, thence eastward over southern New England into the Atlantic, attended throughout its course by heavy rains and brisk high winds, increasing at times to gales. These rains were particularly heavy in northeastern Florida and southeastern Virginia, over 6 inches of water falling during the storm at Jacksonville, and nearly $7\frac{1}{2}$ inches at Norfolk. In the record of averages at the Kew Observatory, England, I find: "Rainfall during the month 6.46 inches, of which 3.9 fell during the seven last days of the month.

I have heretofore stated that strongly marked phenomena frequently occurred midway between the Vulcanian equinoxes, which leads me to suspect that his periodicity really is but 23 days, consequently that one of his equinoxes occurs every $11\frac{1}{2}$ days. The enormous rains and destructive floods in India from the 19th to the 23d of September is one of these phenomena. To be sure it may have been caused by the Telluric equinox of that date. But the general principle that phenomena are composite, that is, the effect of two or more causes gives reasonable ground to suspect that there was another disturbance superimposed on the Telluric. On the 13th, 14th, and 15th of October, 1872, there is almost positive proof of the agency of an intervening cause of disturbance, that has not yet been accounted for. To be sure Mercury passed his equinox on the 9th of October, but it is unusual for him to manifest much energy at a period four or five days removed from his equinox. Besides a storm centre passing over the continent from the 13th to the 16th, and a second one on the 17th, we have large numbers of sunspots —one so large as to be visible with the natural eye—reported by Mr. Denning on the 13th and 14th, with intensely strong earth currents simultaneously in telegraph lines in England and on the continent of Europe, so strong as to interrupt at times all communication. These currents continued until the 18th, and were only intense on lines running East and West; lines running North and South were but little affected. Prof. Sperry, of the Stonehenge Physical Observatory, made particular observations on them from the 14th to the 18th of the month. The oscillations and declinations of the magnetic needle were sudden from East to West, and *vice versa*. The movement of the vertical force was frequently too great to be recorded on the photograph paper, and the magnet was frequently thrown off its balance. Furthermore, while I am writing this (July 22d, 1875) tremendous rains have fallen for the last six days in every part of the American continent, and also in Europe; and at some localities, Wheeling, West Virginia, for instance, destructive tornadoes have occurred. Mercury's equinox occurred on July 18th, and, as expected, heavy rains fell on that day, and two days before and after; but it is unusual for this planet singly to show so much and such a long persistence of energy.

The 14th Vulcanian equinox of 1872 occurred on November 16th. From the 13th to the 17th a storm centre passed slowly over the continent from West to East, reaching the lower St. Lawrence Valley on the 17th. On the 13th, rain or snow is reported at nineteen stations of the Signal Service; on the 14th, at thirty-five stations; on the 15th, at twenty-six stations; on the 16th, at seventeen stations; and on the 17th at twelve stations.

The 15th and last Vulcan equinox of 1872, occurred on December 8th. On the 6th, a storm centre appeared in the extreme northwest; it was central in the Lake Region on the 8th, and reached the Gulf of St. Lawrence on the 10th. Simultaneously with its disappearance, three storm centres appeared, two in the extreme West and Northwest, and one in Mexico. The northern and central ones pursued the usual easterly course over the Lakes and Ohio Valley; the one from Mexico passed over the Gulf and Florida, thence along the Atlantic coast, with the Gulf Stream, towards the northeast.

Fifteen Vulcanian equinoxes occurred in the year 1872, which we have presented seriatim together with the phenomena—the greater part of which are the observations of the Signal Service—proving that at every one of the periods, if not manifested before, then generally on the very day the equinox occurred, a disturbance appeared, which produced marked phenomena; many of them being very violent. As we insert in the Appendix, a table of all the Planetary Equinoxes that have occurred since the 1st of January, 1866, as well as those that will occur until the 31st of December, 1884, so we might safely rest the case here. Every reader by observing the phenomena occurring at the equinoxes that have to come, can verify the theory for himself; and those who have the Annual Reports of the Signal Service, for the years 1873 and 1874, can make a verification by comparing the dates of the equinoxes with those of the current phenomena at the period. But as these reports are not accessible to all, and in the hands of but few, we will have to complete the verification of the Past, trusting to the interest that the reader will take in the subject to make verifications of the Future. We will however pass over as rapidly as possible the review of the phenomena, unless where their intrinsic importance, or extraordinary character, demands particular attention. To avoid so much repetition, we will designate them merely by numbers:

### VULCANIAN EQUINOXES OF 1873.

No. 1 occurred on January 1st. A high barometer from the northwest moved from the 1st to the 4th over the Lakes, southeast. Storm centre No. 1, coming from the Pacific coast, in the rear of this high barometer, hence was deflected on the 3d, from Wyoming into Arkansas, where it was on the 4th. In the meantime, the high barometer having passed into the Atlantic to join the permanent Bermuda high barometer opposite the coast of

North Carolina; storm centre or low barometer No. 1 swept by a curve through Tennessee and Kentucky, thence up the Ohio Valley northeast to the Gulf of St. Lawrence.

No. 2 occurred January 24th. On the morning of the 23d there was a low barometer central at St. Louis; it passed diagonally through Illinois, Indiana and Ohio, to the southern shore of Lake Erie, thence eastward into the Atlantic on the coast of Massachusetts, where it joined, on the 24th, another low barometer from Texas, through the interior of the south Atlantic States. These two storms covered the whole country east of the Rocky Mountains with heavy snow and rainfalls, accompanied with severe gales on the Lakes and the Atlantic coast.

No. 3 occurred February 16th. A low barometer coming from Mexico, was central at the mouth of the Wabash in southeastern Illinois on the morning of the 16th, thence to the Atlantic coast opposite New Jersey. This was followed on the 17th by another low barometer, which passed northeast over Lake Superior. The first was accompanied by very heavy rains and snows in the North: in the South with thunderstorms. That on the 17th was followed by a high barometer and *very low temperature.*

No. 4 occurred March 11th. Of the low barometer that passed over the continent from the 10th to the 11th, the Signal Service Report says: "It passed from Dakota over Minnesota and Lake Superior into Canada, down the St. Lawrence Valley; preceded by brisk and high easterly and southerly winds, and followed by brisk to high westerly to northwesterly winds; accompanied by rain and snow from Missouri to Minnesota, and eastward over the Lakes. In the Eastern and Middle States the rains and snows became general and at places very heavy."

No. 5 occurred April 3d. On the 1st and 2d a storm centre passed over Missouri, northeast into Canada. The Signal Service say it was accompanied by brisk winds and heavy rain in all the States east of the Rocky Mountains. It was felt as a severe storm from Texas to the Lakes and Northwest. Of the low barometer that passed over the continent from the northwest from the 3d to the 7th, the Signal Service say it was accompanied by brisk and occasionally high winds and rain from the Missouri through the Ohio Valley, thence to the East and middle Atlantic coasts, Lakes, and St. Lawrence Valley. The light-house at Erie was blown down during this storm.

No. 6 occurred April 26th. The phenomena of this equinox are complicated with those of the Venusian equinox of April 30th which have already been quoted, which were tremendous heavy snow falls in Kansas and Nebraska, with excessive rains in the South.

No. 7 occurred May 19th. A low barometer, originating on the 17th on the Southwestern Plains, traveled slowly northeastward through the Indian Territory over Lakes Michigan and Huron into Canada, attended with rain and brisk winds. Simultaneous with its disappearance on the 20th, a low barometer appeared in the Northwest. It was central at a point about 100 miles Northwest of St. Paul, at 4 P. M. on the 22d. At this hour it was that on the same meridian in Washington County, Southern Iowa, the terrific and destructive tornado occurred, of which a full account is given in Part I. Mercury's equinox had occurred on May 17th, or two days before the Vulcanian, and five days before this terrible visitation.

No. 8 occurred June 11th. Of a low barometer that passed slowly over the country from Dakota to the New England States, from the 7th to the 11th, the Monthly Weather Review says, "it was accompanied with rain, generally heavy, fresh and brisk winds, and occasionally severe thunderstorms." Two other storm centres, one in the Southern States and the other in the Northwest, appeared on the 12th, and accompanied with the usual phenomena.

No. 9 occurred July 4th. Several storm centres passed over the continent from the 2d to the 5th. On the 3d there was a tornado at Indianapolis; and shocks of an earthquake were felt on the 3d and 5th at Buffalo.

No. 10 occurred July 27th. On that day storm centre 13 of the Signal Service passed from the Northwest over Lake Superior. Among the Lake disasters of this period, schooner *Two Brothers* lost jib-boom and foremast head by a storm. Schooner *Sunrise*, foremast shivered by lightning on Lake Erie. Tug *Ada Allen* sunk, etc.

No. 11 occurred August 19th. We have already exhausted the phenomena of this period in what we have said of the Nova

Scotia cyclone, which at this date was between the Bermuda and Bahama Islands, on its way to the coast of Nova Scotia, which it struck five days later.

No. 12 occurred September 11th. The Monthly Weather Review says of low barometer No. III, "September 11th, 12th and 13th was accompanied by rain at nearly all of the stations east of the Rocky Mountains; brisk and high winds over the Northwest and Upper Lakes; severe "Norther" in Texas on 13th; and heavy snow on Mount Washington."

No. 13 occurred October 4th. There were no less than five low barometers or storm centres developed between the 1st and 6th of October: three in the Northwest, and two in the Gulf and on the coast of the southern Atlantic States. We will only speak of No. I, first recognized as existing on the 3d, some distance southwest of Cuba. Its influence was felt at Galveston and Indianola on the 4th, from the brisk and high northerly winds that were blowing into it at those points.

The steamship *G. W. Clyde*, Captain James Albert Cole, from Galveston, September 30th, experienced a series of easterly gales, with heavy sea, up to October 3d, when the wind, increasing in force, changed to the southwest. The ship was then run to the westward for more sea room, and during the night of October 4th, the unfortunate vessel was struck by the cyclone, her position then being about fifty miles westward of the Tortugas. The steamer was tossed about in the most fearful manner. At 6.40 P. M., on the 5th, the sea swept the pilot-house away, and Capt. Cole, together with four others of the ship's officers, were washed overboard and drowned, with the exception of Quartermaster Burns. The vessel was afterwards towed into Key West.

At this place the barometer had been steadily falling during the 5th, with variable easterly to southeasterly winds. The wind blew a gale at 6 A. M. of the 6th, and continued so, with occasional lulls during the entire day. After 4 P. M. the mercury fell very rapidly, reaching its minimum at 5 P. M., when it had fallen to 29.28. After this it commenced to rise again; but the wind shifting to the southeast, increased to a hurricane at 6.30 P. M., the velocity twice reaching one hundred and twenty miles per hour, as shown by the self-register of the anemometer.

During the passage of the storm, almost the entire island was flooded, the sea rising fourteen feet above mean tide, rushing through the street from two to four feet in depth; and the salt spray which filled the air, destroyed vegetation as completely as a severe frost.* The damage to shipping in the harbor, although prepared for the storm, was immense. On the island, houses trembled and shaked to their foundations; roofs were carried away like straws, but fortunately but one life is reported lost.

At Tortugas, 65 miles west, the wind blew terrific for twelve hours. Colonel Langdon, in command at Fort Jefferson, states that a solid bar of iron, weighing eight hundred pounds, was carried two hundred yards over the parapet of the Fort.

The cyclone, which was progressing northeastwardly at the rate of about five miles an hour, struck Punta Rassa, Florida, about 4 P. M. on the 6th, and its very centre seems to have passed over this point. The barometer here was similarly affected as at Key West, during the 4th and 5th, and the morning of the 6th; but after 4 P. M. it fell more rapidly. At 4 P. M. it stood at 29.10; at 8.45 P. M. it stood at 28.40, being the lowest point it reached.

The scene at this station, while the storm was at its height, is said to have been most terrific. The anemometer being blown away at 4.11 P. M., there was no means of ascertaining the velocity of the wind. The sea rose eleven feet above the mean tide, all land was flooded, even the highest points, leaving the inhabitants entirely without drinking water. Houses and wharves were swept away, cattle were drowned, and the entire loss at the place, including damage to shipping, is reported at $116,400; but no lives lost. The course of this cyclone, after it had forced a passage between the two opposing high barometers, was rapid northeastwardly with the Gulf Stream. Of course, after the passage was effected it lost its cyclonic character. It is supposed to have been the same storm that passed over the Shetland Islands on October 10th.

No. 14 occurred on the 27th of October. A general atmospheric disturbance followed this equinox; no less than three storm centres from the 25th to the 28th, originated or passed

---

*NOTE.—This will be considered an important fact when people once get rid of their mechanical theory in Physics.

over the continent. Low barometer No. XII, of this month, originated in Texas, and was central in the Indian Territory on the 25th. "Snow was then falling throughout the Northwest very heavily, with high winds. The precipitation became very heavy on the 26th, when it passed over the Lakes; with heavy thunderstorms in the South. It was central in New Brunswick on the 27th and 28th, with southern gales in Maine; reported the heaviest storm of the season at New Haven, Wood's Hole, Portland and Eastport." Low barometer No. XIII, October 27th, 28th and 29th. "The snow and fog that prevailed on the Lakes united with the high wind, made the night of the 28th and 29th one of the wildest description, and numerous disasters are reported from Milwaukee."—*Monthly Weather Review.*

No. 15 occurred November 19th. Within three days no less than four low barometers or storm centres originated and passed over the Continent. We can only quote from the Monthly Weather Review a condensed description of No. VII, which was the most remarkable cyclone of which the Signal Service observations furnish any details. It was generated about midday of the 16th, in Northern Georgia, and at once assumed a threatening aspect. During the night of the 16th and morning of the 17th, it steadily advanced to the vicinity of Wilmington, making about 240 miles in 12 hours. Its course thence was northeastward along the inshore margin of the Gulf Stream, which it tenaciously followed to latitude 43° north, whence it struck off into the Bay of Fundy, and thence to the mouth of the St. Lawrence River. All along its track, from Norfolk to Halifax and Father Point, was marked by fierce gales, and vessels report fearful seas off the coast. At Norfolk, chimneys and fences were blown down, and shipping in the harbor dragged anchors. In Chesapeake Bay it was extremely severe. At Cape May the barometer fell to 28.76, with very heavy sea; and pilots from outside report it the worst gale known for years. At New Haven, Wood's Hole, Boston, and Portland, Maine, it fell very low, at the latter point to 28.49, the lowest observed barometers since these points became signal stations. At Eastport, Maine, on the 18th, the cyclone attained terrific force, its winds blowing 64 miles an hour. Its progress over the Canadian districts to the northward and eastward was equally violent.

No. 16, and last of 1873, occurred on the 12th of December. The phenomena of this equinox are complicated with those of the Venusian equinox of December 9th. This month was marked with excessive rains, as is usually the case when a Venusian equinox occurs. The Weather Review says, "No. VI (low barometer) began in New Mexico, and during the night of the 10th, and the forenoon of the 11th, advanced eastward to the central Arkansas Valley, and thence moved northeastwardly over Missouri and Illinois on the 12th. It passed eastward over New York, and reached the Atlantic coast on the night of the 13th, and on the morning of the 14th its centre was near Halifax, Nova Scotia.

We have now come to the close of another year, and in passing under review the dates of the sixteen equinoxes of Vulcan that fell within the year, in not a solitary case have we failed to find cotemporaneous physical phenomena of a very marked character. Not only so, but the most extraordinary phenomena that have occurred during the year, have all fallen upon these Vulcanian periods. Other extrordinary phenomena falling upon these periods were so far removed from any other known cause, that unless they are due to Vulcan's equinoxes, no other cause can be assigned for them. If the theory that all physical phenomena result from astronomical causes, has any foundation in fact—which will hardly be contested while the facts we have adduced stare the doubter in the face—then it must be admitted that unless these sporadic, and oftentimes very violent phenomena, are not attributable to the Vulcanian equinoxes, then no other astronomical cause can be assigned for them.

### VULCANIAN EQUINOXES OF 1874.

No. 1 occurred January 4th. Storm centre No. I, originating on the Pacific, was central in Minnesota on the 3d, and reached the Gulf of St Lawrence on the 5th. It was accompanied by rain and snow, and followed by a high barometer, that, as usual, brought very cold weather. Low barometer No. II appeared in the Gulf on the 5th. It passed into Florida at St. Marks, thence through Western Georgia and North Carolina, it crossed the Alleghany Mountains and over the eastern part of Lake Erie into Canada. The *Weather Review* says: "It was this storm that produced the snow and sleet in the Lake Region and Ohio Valley, on the 6th and 7th of the month, and severed the telegraphic communication between the East and West."

No. 2 occurred January 27th. Low barometer No. VII, of January, appeared at Yankton on the morning of the 27th, passed over the Lake Region during that day, and passed into the Atlantic, north of Boston, on the 28th.

No. 3 occurred February 19th. On the 19th a storm centre, coming from the Pacific coast, passed over the Lake Region, reaching the mouth of the St. Lawrence on the 20th. There was another storm centre that passed over Texas on the 16th to Florida on the 18th; and one on the 20th from the Southwest, through Arkansas to Cairo, Illinois, thence up the Ohio Valley.

No. 4 occurred on March 14th. Low barometer No. VII formed in the Northwest on the 14th. Being both retarded and deflected by a high barometer over the Lakes, it did not reach western Missouri until the 17th. It passed St. Louis, after sweeping over Southern Missouri, on the 18th, thence to Milwaukee, where it reached that evening. The Weather Review says: "It was accompanied by violent lightning, thunder and heavy rain. While the storm centre was in eastern Missouri, and making a northward or northeastward curve in its course, a severe storm or tornado was formed in the Mississippi Valley, which, at 4 A. M. of the 18th, swept with great fury over Cairo, Illinois."

No. 5 occurred April 6th. Low barometer No. III developed on the 4th, in Texas. Its course was due northeastward over Lake Erie, thence to the Gulf of St. Lawrence, which it reached on the 6th. "Very extensive rains uniformly prevailed over the Southern and Atlantic States, and snow over the Lake Region and Upper Mississippi Valley."

No. 6 occurred April 29th. Low barometer No. XII, originating in the mountains of Colorado, on the 25th and 26th, but retarded by high barometers in the Lakes, passed through southern Missouri on the 27th, thence eastward south of the Ohio, to Albemarle Sound, which it reached on the 28th. The Review says: "Heavy rains prevailed on the night of the 27th–28th in the Ohio and Tennessee Valleys and snow on the Lower Lakes.

When it struck the Atlantic it turned sharply to the northeast, passing over Maine on the 29th, at midnight."

Of low barometer No. XIII, the Review says: "The history of this remarkable storm belongs especially to the month of May; during the first five days it moved slowly from the northwest to Tennessee, thence to Cape Hatteras. The origin of this storm was on the Pacific coast. It passed over Washington Territory on the 28th and 29th, and over Montana on the 30th. Earth currents are reported at Pembina (pronounced Pem-bin-aw) on the 29th, the day of the equinox

. No. 7 occurred May 22d. Storm centre No. VI passed over Lake Erie on the 20th; New York, 21st; and Maine, 22d. Of low barometer No. VII, which appeared in Dakota on the 22d, the Review says: "It was accompanied on its north and east sides by an unusual number of thunder and hailstorms, of which a special report is in preparation." These thunderstorms, occurring on the afternoon of the 25th, are reported from nearly every station in the Gulf States, Tennessee, Virginia, middle Atlantic coast, Lake Ontario and New England.

No. 8 occurred June 14th. Of low barometer No. VI, the Review says. "It passed from Kansas on the 14th, northeastward over the Upper Lakes, thence eastward and southeastward over Canada and New England. The barometric gradients of this storm centre were, without exception, the steepest recorded during the month." It gave rise to remarkable storms over the whole country.

No. 9 occurred July 7th. Low barometers III, IV and V, the first two coming from the West, the other originating in southern Illinois, passed over the country from the 3d to the 10th. Tornadoes occurred on the evening of the 4th, in Pennsylvania, Maryland, Virginia, Delaware and New Jersey; and on the 7th at Three Mile Bay, New York. Earthquake on the 9th at Cairo.

No. 10 occurred July 30th. Low barometer No. I, of August, originated in Dakota on the 29th of July; it was on the Upper Lakes on the 30th, and the Lower on the 31st. The phenomena of this equinox are complicated with those of the equinox of Venus on July 23d, and are already given—while the Venusian cycle was under discussion—in the account of the tremendous water-spouts during the last week of July, 1874.

No. 11 occurred August 22d. Storm centre No. VII brought rainy weather to all States east of the Rocky Mountains; it orignated in the Northwest on the 20th; it divided into two branches, one proceeding east over the Lakes, the other by the mouth of the Ohio to Cape Hatteras; severe thunderstorms accompanied both of its branches across the Continent.

No. 12 occurred September 14th. As a Mercurial equinox occurred on this same day, the cotemporaneous phenomena have already been given in the discussion of the Mercurial Cycle.

No. 13 occurred October 7th. Storm centre No. III passed from southwestern Texas, through the heart of the country northeastwardly to the Gulf of St. Lawrence, from the 7th to the 9th. It was preceded on the 6th by storm centre No. II from the Northwest. The two centres formed a junction in Central Pennsylvania.

No. 14 occurred October 30th. Low barometer No. IX appeared in southern Colorado on the 27th, passing centrally over the Lake Region, it reached the Gulf of St. Lawrence on the 30th. This was a storm of great severity, marked with nearly a straight track, and accompanied with high cyclonic winds. At Duluth the sea on Lake Superior was very violent. The rocks on the outside of the breakwater were thrown on the inside. Several vessels coming into port from the north shore of the Lake, were disabled, and one lost.

No. 15 occurred November 22d. On the 20th, low barometer No. VIII appeared in western Wyoming. This proved to be an extraordinarily violent and extensive cyclone. It passed through North Missouri in the forenoon of the 22d, and reached Chicago at night. It was on the 22d, the day of the equinox, that a terrific tornado struck the town of Tuscumbia, Alabama, destroying upwards of 100 buildings, and killing 12 persons. Nearly half of the town was destroyed, and an immense amount of damage done otherwise. Almost simultaneously the town of Montevallo, about 60 miles north of Selma, Alabama, shared a similar fate. The electric phenomena of this storm were very extensive, intense and unusual, since the cold season was already far advanced.

No. 16, and last Vulcanian equinox of 1874, occurred December 15th. Storm centre No. VI appeared in the Northwest on the 12th, and reached the Bay of Funday on the 14th. It was very severe, and increased in energy as it approached the coast. Several wrecks occurred on the Atlantic coast, north of Cape Hatteras. A decidedly very high barometer followed immediately in the rear of this storm centre, accompanied by very cold clear weather.

In our investigation we have traversed the circuit of another year, and again we have found not only a uniform but an invariable correspondence of dates between physical phenomena and the Vulcanian equinoxes. Moreover the facts show that the most extraordinary one during the year occurred on the very day of one of these equinoxes, and all its strongly marked phenomena crystalize around other equinoxes. Such invariable concurrence is not within the range of probabilities, and there is no way of accounting for it except upon the hypothesis that it is the effect of a fixed and invariable cause.

### VULCANIAN EQUINOXES OF 1875.

The first occurred on January 7th. We have already given its phenomena in full, while discussing the subject of earth-currents. The second occurred January 30th. Low barometer No. XI passed north of the Lake Region on the 29th and 30th. Strong gales along the Lakes, especially at Toledo on the 30th. It was accompanied with snow and brisk to high winds. Low barometer No. XII appeared in the Northwest on the 29th, passed centrally over Illinois and Indiana, thence over Lakes Erie and Ontario. Another low barometer developed in Georgia on the 30th, which latter became a severe snowstorm in New England. In Texas and the Gulf States it produced thunderstorms, north of Tennessee, snow and sleet. An aurora seen on the 31st.

The third occurred on February 22d. Low barometer No. XI appeared in the Northwest on the 21st. It passed eastward to Ohio on the 22d, where it seems to have propagated its influence to a depression in the Indian Territory, during the night, and on the 23d it moved northeastward into Iowa. It was during its passage, on the 23d, through Missouri, that the town of Houstonia was destroyed by a terrible tornado.

The fourth occurred March 17th. Low barometer No. VIII formed in Texas on the 17th; on the 18th and 19th it moved slowly into Louisiana; on the 20th it developed into the most terrific tornadoes from Central Alabama, over Georgia, into South Carolina. Here I cannot refrain from entering a most earnest protest against the culpable neglect of what are ostensibly scientific associations and institutions, established and endowed to promote the increase and diffusion of knowledge amongst Mankind, in not giving a thorough investigation to these rare and most significant phenomena.

The economy of Nature is conducted not only upon the most simple principles, but by the most efficient, because omnipotent, causes, and the most stringent and effective laws. The cause producing a low barometer does its work efficiently, and with a power that surpasses all human calculation. It draws to the centre of continents moist air from all the surrounding oceans, though thousands of miles away, and sends it to the upper regions of the atmosphere as a volcano does its smoke, ashes and scoriae, or an immense fire the superincumbent heated air, there to form clouds to distil and distribute the grateful rain as they are borne along by currents of air going to be poured down again through the ærial maelstrom over a high barometer. Who can estimate the amount of work it performs? Look at the mighty rivers that perennially flow into the sea! What is the weight of the water they annually bear back and pour into the ocean? What power is it that conveys these waters from the ocean to the highlands and mountains in the centre of continents where rivers have their sources? To St. Louis, in the heart of the Mississippi Valley, sufficient water is annually carried to cover the surface of the Earth nearly four feet in depth, and in some exceptional year, it has nearly touched six feet. What is the weight of this immense body of water? St. Louis is fully a thousand miles from the nearest ocean. What then is the force that transports and the motive power that puts this immense body of water in motion, and conveys it 1,000 miles into the Continent? What answer has Science—falsely so called—to give these questions? None; absolutely none whatever; for the hypothesis that it is Heat does not bear the least scrutiny. If it be Heat, then this upheaval of air would take place over the hottest portions of continents, which is never the case. In America it would take place in the Gulf States; in Africa, in the

Great Sahara; and in Asia, in Arabia, and India, which every tyro knows is not the case. In America it is in the Rocky Mountains, and especially in that part of them in western Manitoba, at the sources of the Missouri and the Sasgatchawan Rivers, and north of the forty-ninth parallel of north latitude. In Asia it in the Altai Range, on the southern border of Siberia, where the Obi, Yenesei, Lena and Amoor have their sources; and in Africa it is over the mountains of South Africa, at the springs of the Nile, the Congo, the Orange and the Zambezi. Then, again, if the thermal hypothesis be true, the heaviest precipitation would take place at the hottest season of the year, which again is known not to be the case. We get the heaviest rains first, and the hottest season afterwards. According to the thermal hypothesis, the effect then precedes the cause. A theory that ignores such positive facts, and is so oblivious of all logical deductions—that is, of Common Sense—requires no refutation.

Here, in this transportation of such a vast amount of Matter, a motive power is operating, millions of times greater than the power of all the horses in the world, together with all the mechanical powers Man has contrived or can command, still, not yet has he discovered the nature and character of the power. In fact, he virtually ignores it. He talks flippantly of indrafts of dry cold air on one side, and of moist warm air on the other, and then, incoherently, of conflicts of these counter currents in which hurricanes are born. These phrases are full of sound and fury, but absolutely "signifying nothing."

Indrafts! Then there must be motion. Now, what is the cause of this motion? Let that which claims to be Science, answer if it can. The cause of motion! Why; is it necessary for the Stagyrite to arise from the dead and repeat the lesson he taught twenty-three centuries ago: "*That all that moves leads us back to the cause of the motion that we perceive.*" It certainly does seem so; for here we have *motion*, but what passes for Science in the Nineteenth Century, has not ascertained the *cause* of that motion. This cause works effectively, but so silently and inobtrusively that Man has not perceived and recognized it, though it is as open as day. In the tornado it assumes a visible embodiment and acts so terrifically as to inspire Man with consternation and awe. There that cause stands revealed as Electricity in its utmost tension, and there it discloses the laws by which it operates; but our institutions, who have charged themselves with the task of increasing and diffusing Knowledge amongst Mankind, culpably neglect and studiously avoid an examination of all the attendant circumstances and characteristic facts of these rare, terrific, but significant phenomena. Why? Because indrafts of air and conflicts of atmospheric currents are too insignificant to demand attention from the high priests of Science. In this we would concur, looking at the matter from their stand-point, which is that these indrafts and conflicts are accidental, and happen without an efficient cause.

Let us, however, be just to them. They do sometimes investigate them, but it is by one who has nothing to learn. He finds nothing, because he is in that state contemplated by Plato, when he asks: "How can we expect to find unless we know what we are looking for?" Looking for indrafts, he easily finds them. But he is not, nor does he make us any wiser than we were before. We knew before, that where there is such an *updraft* there must be a corresponding *indraft*. We know that a low barometer is fed by surrounding high barometers; and therefore it is a natural inference that an arctic high barometer must pour into such a low barometer a supply of dry cold air; and that an oceanic and subtropical high barometer must pour into it moist warm air. The conflict amounts only to this: it is the collision of warm air with warm air, as well as warm air with cold air in the rush into the cyclonal vortex. But what we want to know is the *cause* of this vortex, and the laws and modes of its action. This we can only ascertain by a minute and exhaustive scrutiny of the facts in all their details and integrity; and these facts can only be ascertained by an intelligent, full and discriminative investigation of phenomena as soon after their occurrence as possible.

The fifth Vulcanian equinox occurred April 9th. On the 8th low barometer No. III passed through Nebraska with heavy rainfall in the Missouri Valley, into Northwestern Iowa, then centrally through Minnesota, northwest of Lake Superior. No. IV appeared in the Southwest on the 10th, passing through Arkansas, thence up the Valley of the Ohio on the 10th and 11th, with high winds, heavy rains and snows, and with severe thunderstorms in the Southern States.

The sixth occurred May 2d. Low barometer No. I on the 1st and 2d, passed from the Ohio Valley northeastward over the Lower Lakes, accompanied by high winds and heavy rains and snows. Local tornadoes in Georgia, South Carolina, Tennessee and Indian Territory, and destructive gales on the Lakes and Atlantic coast. Storm centre No. II followed from Kansas on May 2d, and No. III on the 4th. The latter giving rise to severe hailstorms. Earthquakes occurred in Asia on the 3d and 4th of May.

The seventh occurred May 25th. Low barometer No. X developed in the extreme Northwest on the 21st. It moved slowly into the Mississippi Valley, and covering it with a system of depressions, some of which developed into tornadoes in the Indian Territory. On the 26th it developed into one centre in western Nebraska and Dakota, and then passed eastward over the Lakes.

The eighth occurred June 17th. Low barometer No. V of the Signal Office for June, passed over the Mississippi Valley on the 15th and 16th. At Quincy on the 15th there was a severe tornado and tremendous rain, causing destructive floods. The same was the case at Hannibal, Missouri. Heavy rains extended West into Kansas, and Northwest into Nebraska. On the 16th, 17th and 18th, the rainfall was enormous, and the winds terrific. The same was the case over all the country east of the Rocky Mountains, but especially so in the Missouri, Central Mississippi and Ohio Valleys; destructive tornadoes occurring at many places. In Indiana, Ohio, and northward, an earthquake was felt on the 18th, at many points, while a violent storm was raging. We have information that at McConnellsburgh, Fulton County, Pennsylvania, a tremendous rain, hail and wind storm passed over that section of country on the 18th. The influence of this equinox became complicated with that of Venus, which happened eight days later. Europe felt the full force of the perturbation in the tremendous rainfalls and destructive floods in France on the Garonne, and in Hungary. Heavy rains fell in California on the 14th and 15th of June, unusual at that season; Chili had tremendous rainfalls, and at Valparaiso a destructive storm and flood. New Zealand had immense and unusual rains, and a violent cyclone. China had her typhoons. All parts of the Globe in fact were involved, as necessarily must be the case under the operation of a cosmical cause. The intense severity of the paroxysm that followed two or three days after the Venusian equinox makes me suspect more strongly than before, that Vulcanian equinoxes occur every $11\frac{1}{2}$ days, and that his periodicity is only 23 days. The Mercurial equinox of July 18th, 1875, just passed, strongly corroborates this view, as already stated, by its extreme energy and prolongation. There is a phenomena connected with the Mercurial equinox of July 18th, 1875, worthy of a record: The 17th at St. Louis had been unusually warm. At sunset a cloud appeared in the northern horizon, extending from northeast to nearly west, which, as night approached, proved to be an extended thunderstorm, the flashes of the lightning along its whole line, were brilliant, beautiful, and almost incessant. I am informed a heavy storm raged at Normal, Illinois, at that time. At about 10 o'clock, P. M., a heavy mass of scud cloud separated from the thunder-cloud and drifted rapidly due South, accompanied by a brisk and extremely chilly wind. Sergeant George Prender, in charge of the Signal Station, informs me the thermometer fell seven degrees in five minutes, which is almost as rapidly as it would fall if placed in ice. Such phenomena have a revelation to make, when we once know how to interrogate them.

The *New Orleans Picayune* of 20th of July, in noting the phenomena of the prevailing Mercurial perturbation, and calling public attention to their extraordinary character, says: "The change of the 18th was of the most radical and extraordinary character. The heat of the previous day had been almost unbearable—a heat rare and exceptional in New Orleans. The 18th dawned cold, raw and shivering; such a day as we would have grumbled at in November." It gives the rainfalls at New Orleans of the night of the 18th, and morning of the 19th, at seven-tenths of an inch. It hence appears that what I supposed at the time to be only an insignificant, local and temporary eddy, was in fact a great atmospheric movement from the polar towards the equatorial region, traveling with great velocity almost due South.

The ninth and last Vulcanian equinox that occurred when this volume was put to press, was that of July 10th, 1875. A low barometer moved over the Northwest, the Upper Lakes and Canada, on the 9th, 10th and 11th. On the 10th cloudy and rainy weather covered the country from Missouri and Arkansas east to Virginia and New Jersey. Heavy rains, local floods, and in some localities severe tornadoes continued to occur for four and five days, when there was a short intermission. Heavy rains and floods were again reported in Europe, especially in France and Austria. A series of damaging local floods also occurred in England, which, at the Mercurial equinox of the 18th, became very extensive and destructive.

The following telegram, giving a description of an awful hail-storm on July 7th, 1875, is one of the attendant phenomena of this—Vulcanian—equinox. Besides sustaining the theory, it is too important an account of the facts it states to be omitted:

"A correspondent* at Geneva gives an account of the terrible storm which broke over that city at midnight of Wednesday, the 7th inst. About eight o'clock a few heavy drops of rain began to fall, and at the same time the whole circumference of the horizon began to be fitfully illuminated by flashes of sheet lightning, but there was no thunder. Once only during the entire night was there one terrific clap of thunder, and that was when the storm was just over. The lightning gradually increased in intensity, and became indeed actually and without exaggeration continuous. The entire atmosphere seemed to be an element of flames, and all this time there was not the slightest movement of air in the streets. At about eleven o'clock small objects lying on the roofs of the houses were caught up and whirled around as by a cyclone. Still there was no movement of air in the street below. At midnight the tempest came mainly from the Jura range and from southwest, traveling in that direction toward the basin of Lake Leman. As it neared the lake it seemed to spread out into a fan-like form, with a front sufficiently wide to embrace the entire city; it did not last much longer than ten minutes. At the end of it Geneva was wrecked as no army of besiegers could have wrecked it in the same space of time. The storm came in the shape of an almost compact mass—a sheet of ice, driven horizontally before the tempest blast. In the first

---

*Note.—T. A. Trollope, of the *N. Y. Tribune*.

instant every gas lamp in the streets—save here and there one was spared by reason of some protecting roof—was smashed to atoms and extinguished, but the city was not in darkness, for the masses of coagulated hail reflected the blue light of the lightning in a ghastly and ominous manner. The windows of manufactories and residences were forced from their fastenings, besides having all the glass shivered, and bedrooms and staircases and saloons were thus thrown open to the storm, and in a minute or two half filled with masses of ice far beyond the immediate power of the inhabitants to remove, for the storm was marked by the peculiarity that the hailstones, or ice fragments rather, compacted themselves into a solid mass as soon as they fell. On the slope of the left bank of the river Arne, in the suburbs of the city, the tiles of many houses were absolutely beaten to powder; stout partitions of wooden plank were pierced by holes such as might have been made by a musket ball. Three persons were killed by the fall of a farm house in the immediate vicinity of the city. Vast quantities of small birds have been picked up, killed by the storm, and the bodies of several foxes have been found. Geneva is surrounded by pleasure gardens, vineyards, and market gardens, and these having been destroyed as if a charge of cavalry had passed over them, involves the ruin and despair of poor and industrious peasants, whose all is now taken from them as effectually as if it had been sunk to the bottom of the sea."

The sheet lightning, without thunder, is the same that Mr. Wise and other æronauts have seen passing between the lower and upper cloud discs of a thunderstorm, of which we have fully spoken in Part I. The manner in which hail is formed, or rather the cause and law of hail formation, we also have fully discussed there; and the consolidation of the hail after falling, is evidence of the extreme low temperature prevailing where the hail was formed. The facts here stated prove the cause we there assigned for the low temperature, to be the true one. The blue lightning! This we also discussed. *It is the characteristic color of the Electricity at the negative pole.* Small objects lying on the roofs of houses caught up and whirled around as by a cyclone, when there was not the slightest movement of air in the streets! Why, is not this the same phenomenon as the dancing figures on an electrified disc? This is not only a very graphic and interesting description of such a storm, but it presents more important facts that any other we have met with before. But we explained and demonstrated the cause and law of all these phenomena long ago.

The verification of the Theory of Planetary Meteorological Cycles must rest here as far as the Past is concerned; not because the material is exhausted, but for want of time and room; for I have not presented one tithe of the facts I have collected to test and establish the truth of the Theory to my own satisfaction. Besides, it is not necessary to spend my time in going back and exhuming the Dead Past; for I expect all readers to have become interested enough in the subject to test it to their own satisfaction, by the best of all tests, the coming events of the Living Future. That they may do so, I supply them with all the necessary facilities, so that it is only for them to note the times and make their own observations. With these remarks I leave the Theory to the impartial judgment of all fair minded and candid persons, and will abide by any verdict they may pronounce upon it.

# CHAPTER III.

*General Principles on which all phenomena depend. Hyemal phenomena. Conclusion.*

Incidentally, as great questions have arisen, we have laid down and explained the general principles involved; and hence it remains only to present them in a connected form, so that their relation and dependence can be seen and comprehended at a glance.

Electricity is the ultimate cause of all physical phenomena; hence, whether the phenomena are included in Meteorology or not, as they are all produced in accordance with electric laws, they are explainable, and consequently can only be explained by those laws and principles.

Electricity is a polar force, that is, it consists of two opposite polarities, which are idio-repulsive, but mutually attractive. These properties of Electricity is the basis upon which the Universe is founded. When these two polarities come into direct communication, they instantly obliterate each other. It is the struggle of each to free itself, and rush and embrace the other, and the resistance interposed by the law of conservation and compensation, to prevent such a union, that causes the perpetual motion of the Universe. The labor of one is the rolling stone of Sisyphus; of the other, the whirling wheel of Ixion. The achievement of the task is an impossibility; hence the labor towards its attainment is perpetual.

In general terms, our theory of Cosmology may be stated as follows: All bodies in Space are negatively electrified; hence they mutually repel each other, as well as mutually attract each other. There are points in Space where the mutual attraction and repulsion between two, three or any other number of bodies, are *in equilibrio*. These points within limits are the spaces occupied by each planet or sun and the paths they pursue around a central body, or around a point far off in the depths of Space. Space itself is Positive, hence, since all bodies are Negative, Space attracts them. A solar system occupying definite dimensions in Space, will, to a limited extent, neutralize its positive charge, hence render it *quasi* Negative to the Positive in unoccupied Space. The Positive in exterior, or unoccupied Space, will hence attract the solar system; and the *quasi* Negative in occupied Space, and that which has been occupied, upon the same principle, will repel the solar system in the direction of attraction, since it is the line of least resistance. Hence, from the depths of Space untraversed by a solar system comes omnipotent attraction, and from that portion traversed, omnipotent repulsion. This is the cause of the perpetual motion of all solar systems through Space; and a mere modification of it gives us planetary motion around the Sun. But this is forbidden ground for me to traverse in the present discussion; still I have deemed it pertinent to state the general principles, because the perpetual motion of the Atmosphere involves the same motive power in kind, but not in degree.

Electricity being a duple force is duple in its action, that is, all electric action is by couples, one part of the couple being Positive, the other Negative. Not only does static Electricity, that is, Electricity at rest, by Induction evoke its counterpart, the dynamic, that is, Electricity in motion, but the dynamic flowing in any direction by Induction, evokes a counter current to flow in the opposite direction. If the first current be Positive, then the induced or second current will be Negative, and *vice versa*. The second current will induce a third also of opposite Electricity to itself, and flowing in an opposite direction; and so on alternately and indefinitely, each successive current however diminishing in energy.

In Part I we have demonstrated that high and low barometers are parallel electric currents flowing in opposite directions. According to electric laws a low barometer therefore must induce a high barometer, and *vice versa*. Hence it is invariably found that no low barometer ever exists without a high barometer in its vicinage; especially is this the case with low barometers carrying a tropical hurricane.

In that case there is not only in the vicinage of the cyclone a high barometer, but generally a very high one; for the cyclone requires an immense amount of air to supply its upheaving column. By vicinage we mean a radius between, say, from 400 to 600 miles; for in Nature the Cosmical Forces exert their influences through distances so immeasurably great, that distance as measured by Man is too insignificant to be taken into account. Hence we speak of Mercury as being in the vicinage of the Sun, though over thirty millions of miles distant from it; and hence with propriety we can speak of a high barometer as being in the vicinage of a low one, though 500 miles distant. It is hardly necessary to state,—because it is self-evident,—the closer the proximity of the high barometer, the more terrific is the energy of the cyclone.

A high barometer is a descending, and a low barometer an ascending current of Electricity, each composed of electrified air that gives convection to Electricity; one from the surface of the Earth upward, and the other from the surface of the ærial ocean downward. Now since we have shown that the air poured down and out of the anticyclonal column of a high barometer, flows on the surface of the Earth into the nearest cyclonal column of a low one, to be upheaved again, hence this surface current is virtually an electric current, and consequently, according to the law of Induction, it must evoke a counter current in or on the surface of the Atmosphere, flowing from the up-pour of the low barometer to the down-pour of the high barometer. This is not a mere deduction of Reason, but it is a fact established by the observations of the Signal Service. Faint cirrus clouds,—the remnants of exhausted storms that had been generated over the upheaving column of a low barometer,—are seen drifting into the down-pouring vortex of a high barometer. An elongated parallelogram therefore graphically represents the electric couple composing a high and a low barometer, two sides, the down-pouring column and that on the surface of the Earth are Positive; and two sides that of the uprising column and that flowing on the surface of the ærial ocean are Negative. Two points on this parallelogram are static, namely, the base of the down-pouring column under a high barometer and the apex of the uprising column under a low one; and two points are dynamic, namely, the base of the upheaving column under a low barometer, and the summit of the down-pouring column of a high barometer. Each of these points again, if scrutinized, is found to be an electric couple, that is, its action is duple. It has a current flowing to it as well as from it, therefore it must attract one current and repel the other. Hence it is static to the current it attracts, and dynamic to the current it repels. If this parallelogram is still further examined, one half of it, namely, that part within the low barometer, is found to be under a clouded sky; and the other half, that under the high barometer, is under a clear or clearing sky.

High and low barometers are continuous phenomena. We here speak not of the permanent high barometers covering the Oceans on the polar sides of the Tropics; nor of the permanent low barometers of Iceland, of the Aluetian Archipelago, or of that covering the polar regions, but of the transient high and low barometers, the agents that Nature has appointed to distribute in due proportion rain and sunshine over the land so that the Earth will teem with its fruits in their seasons for the sustenance of Man and the lower animals. These are continuous phenomena, that alternately roll from West to East over all parts of the Globe. The duration of their transits vary in length according to circumstances, or according to the electric condition of the Earth and the Atmosphere. In Winter, when the static condition prevails over continents, the tendency for some time, is to form an unbroken series of high barometers in rapid succession. In such case there is a protracted spell of a cold weather. In Summer, when the dynamic state prevails, the tendency is to produce a series of pulsations of low barometers, during whose continuance, there is an influx of a vast amount of moist air from the surrounding Ocean, and consequently of excessive rains. With these exceptional cases, high and low barometers are regular phenomena, following each other in rapid succession at all seasons of the year, but with varying energy. When the Earth and the Atmosphere are free from extraneous influences in consequences of planetary equinoxes, the phenomena attending these respective barometers, such as change of temperature, rain and winds, are mild in form, moderate in quantity, and not extraordinary in energy. But when the electric condition of the

Earth and the Atmosphere is exacerbated by the occurrence of a planetary equinox, then the oscillations of the barometer not only become more frequent, but have greater range. Consequently the phenomena attending them, also become more frequent and manifest greater energy. The Summer phenomena in consequence of the dynamic State prevalent over continents being energized, there is a stronger and more persistent indraft of air demanded, and consequently there is a larger influx of moist air from the surrounding Ocean. The result is a heavier rainfall always takes place in Summer than in Winter; and in cases of intense and persistent energy, the rainfall becomes excessive. When at any period, either in Summer or Winter, the dynamic state has an undue predominance, precipitation is abnormally in excess. Observations in all parts of the Globe confirm this deduction.

The dynamic state heaves up the superincumbent air over a continent, hence necessitates the influx of air from the Ocean. On the contrary, the static condition brings down a maelstrom of air from the surface of the ærial ocean that covers the Earth to an unknown depth. Hence during its prevalence there is an outflow of air upon the ocean; and in Winter not only cold but often intensely cold weather prevails during its continuance.

Now, since planetary equinoxes intensify the electric state, whether it be static or dynamic, consequently it intensifies the phenomena that are offsprings of the given state. In Winter, under such conditions, the downpouring current being energized, the temperature ranges unusually low, especially if the downpouring column advances, as it does nine times in ten, from the circumpolar regions. When, however, the dynamic is unseasonably protracted during any time in Winter, the weather is mild and very wet. But if the energy of a planetary perturbation is mostly expended, as it generally is in Winter, in producing, intensifying and protracting the static condition, then excessive cold prevails during its continuance. Our task now will be to substantiate the truth of this latter proposition by facts ascertained in investigating as far as we had the opportunity and means, the hyemal phenomena attending planetary equinoxes.

In our verification of planetary cycles, it was found that, in America, where the low barometers generally originate on the

Rocky Mountains in the West, they coincide within very narrow limits with the periods of the equinoxes. Consequently, as a low barometer is only one part of an electric couple, its fellow, the other part, must either precede or succeed it in close proximity; or it must do both. In Summer, when the dynamic prevails on continents, and consequently low barometers; the term of the static for producing high barometers is of short duration. Though it brings the refreshing, exhilarating and often bracing west wind so grateful after the sweltering heat that accompanies a low barometer, yet we take so little note of it that we have never inquired into its cause. But the Winter phenomena of the static affect us so disagreeably as to compel attention. The intensely cold weather they bring not only makes us uncomfortable, but exerts a deleterious effect upon domestic animals and upon our fields, orchards and vineyards. These phenomena are fierce snowstorms in the afterpart and rear of the retreating low barometer, followed by arctic cold weather.

The first point necessary to be established is that, intensely cold weather is an accompaniment of planetary disturbances in Winter; and after that, to determine as near as possible the approximate time relatively to the date of the equinox that these cold spells make their appearance, and the order in which the phenomena follow each other. The general order—as far as I am able to determine—appears to be: (1) Five or six days, and even more, before the occurrence of the equinox, an energetic high barometer, and consequently very cold weather; (2) A low barometer on, or a day or two before and sometimes after the equinox, accompanied with moderate, often warm weather, and rain and snow; and, (3) a high barometer with fierce wind and snow storm from the Northwest or West, culminating in intensely cold weather. The cold weather of January 1st, 1864—the most terrible frost ever experienced in the Mississippi Valley—affords a general illustration of the principle here stated.

In St. Louis County the lowest temperature observed was between 3 and 4 o'clock on the morning of the 1st of January, when the thermometer stood $28°$ below zero. At my residence at 5 o'clock A. M., it stood $26°$ below. Dr. Englemann's observation in the heart of the city, at 7 A. M., was $22°.5$. The following were the astronomical condition at the time: Mercury's equinox, December 23d, 1863; Vulcan's, December 28th, 1863, and January 20th 1864; Mars, January 20th, same day as Vulcan; and Venus, February 6th. The weather from Christmas to the forenoon of December 31st, 1863, had been mild, some days quite so, with rain and snow. About the middle of the forenoon of that day, the wind veered to the West, and a furious snowstorm set in, with rapidly falling thermometer; at 4 o'clock the thermometer stood at 12° below zero, with a terrible arctic snowstorm raging that no human being could face. Many persons out hunting, and even on the highway, became bewildered, lost, and perished with the cold. The next morning and day was the coldest ever experienced in the centre of the Mississippi Valley. It extended to the Gulf States. The armies suffered terribly; many soldiers were frost-bitten, and in the Union Army sentinels froze at their posts, in the States of Mississippi, Alabama and Georgia.

We wish we had the daily observations, so as to trace both the range of the barometer and thermometer during this period; but we have only averages, which tell nothing we want to know. Mr. Huron Burt, of Williamsburg, Missouri, in his fragmentary journal, before us, stops at the 19th of December, 1863, and it is not resumed until the 1st of January, 1864. His journal shows rain sleet and snow, up to the 18th. On the 19th he says: "Thermometer at zero, with a terribly fierce wind from the West." This was in consequence of a high barometer following the low barometer that had scattered rain, sleet and snow on the previous days; and this high barometer preceded the Mercurial equinox by four days. What the phenomena were at the equinox, I cannot say, because I can find no *data* that show; and for the same reason I cannot say what they were between that and the Vulcanian equinox five days later. Generally speaking, the weather was mild, with rain, sleet and some snow at the Vulcanian equinox. The high barometer that succeeded the low which had passed two or three days before, brought with it that exceptionally and terribly fierce, cold weather that prevailed for two weeks afterwards, with temperature at zero to 26° below. If the history of this high barometer could be written, it would be found to have originated in the Arctic Circle, somewhere near the mouth of Mackenzie's River on the Polar Sea, and that its path was southeast, passing slowly and centrally over Dakota, Nebraska, Kansas and Missouri, thence through Georgia, into the Atlantic, where it joined the Sargasso permanent high barometer.

As the Martial and Vulcanian equinoxes were approaching, both taking place on the 20th of January, the weather moderated, which is the same as saying the barometer fell and the thermometer rose, for the mercurial columns in the two instruments, except in very rare and exceptional cases, always move in opposite directions.

On the 23d of January, Mr. Burt has this entry upon his journal: "For the last ten days the barometer has ranged from 8° to 30° at sunrise, and from 30° to 52° in the warmest part of the day." On the 29th he makes this record: "For the last week the thermometer has ranged at sunrise from 40° to 60°, and in the warmest part of the day from 69° to 72°. It rained last night."

Dr. Englemann, in the remarks appended to his table of averages for the month of January, 1864, speaking of temperature, says, "the highest was in the afternoon of January 27th; and we had two thunderstorms in one day, on the 29th."

An equinox of Mercury occurred on the 5th of February, one of Venus on the 6th, and another of Vulcan on the 12th of February. We have been unable to find any daily record of the condition of the weather at this time. Mr. Burt gives a summary as follows: "February 16th, after two weeks of unusually mild winter weather, it turned suddenly cold, thermometer at zero. February 18th, thermometer 1° below zero; ground dry and dusty." Dr. Englemann's averages show an unusually small amount of precipitation; atmospheric pressure a small fraction above the general average; the mean temperature nearly 2° higher than the general average; and the principal winds from the West. Wind from the West indicates a high barometer nearly Southwest, that is, south of a line drawn from St. Louis to Santa Fe. A high barometer always draws its supply of air, which it pours out upon the surface of the Earth, from the direction whence it comes. If a high barometer therefore approaches the parallel of latitude of any locality from the South, then it draws its supply of air from the South, and consequently the weather is warm. In Part I we have shown that the high barometer that pours down the simooms felt in Colorado, Kansas, Nebraska, and often in Iowa and Missouri, is a high barometer that has come from the South, and draws its supply of air from the heated plains of the Oronoco and Apure, in South America. Likewise have we there shown that the Arctic cold that overflows the country east of the Rocky Mountains, is the downpour and outpour of a high barometer that has come from within the Arctic Circle, and draws its supply of air thence. The direction that these high barometers coming from the North move, is Southeast, to join the permanent high barometer in the Atlantic, near the Bermudas. The southern barometers I do not consider as detached barometers; but the one that appears on our Southwestern Plains is either the Atlantic permanent high barometer, on an extreme western oscillation, or the Pacific high barometer located west of the Gulf of California, on an extreme eastern oscillation. It is a well established fact that these barometers not only frequently join hands across a continent, but meet and embrace. The Atlantic high barometer, in its eastern swing, frequently covers the whole of Northern Africa.

In America the transient, or wandering high barometers, have three well-defined points, whence they issue from the Arctic Circle; the first is near the mouth of the Mackenzie River; the second near the mouth of the Coppermine; the third about the mouth of Big Fish River. Those originating at the first point, unless deflected by low barometers in their front, ascend the Valley of the Mackenzie into the Missouri Valley, thence to the coast of Georgia and South Carolina; those of the second point ascend westward of Hudson's Bay, over the Great Lakes to the coast of the Middle Atlantic States, they cause the North and Northeast cold winds of the Mississippi Valley, and the Northwest and West cold winds of the New England States; those of the third point ascend east of Hudson's Bay over Labrador, and pour over the New England States those intensely cold north and northeast winds. In Europe they issue from the Polar Ocean, east of the White Sea, and pour those destructive cold north and northeast winds over that continent.

Our inference then is that a high barometer prevailed during the latter half of January and early part of February, 1864, with the exception of about the 28th and 29th of January, when there must have been a low barometer, but that this high barometer was a Southern and not a Northern high barometer. Hence its extreme mildness and dryness; that about the time of the Vulcanian equinox a low barometer passed over the continent, and that an Arctic high barometer followed in its rear, bringing on the sudden change recorded by Mr. Burt on the 16th.

Since energetic high as well as low barometers are characteristic phenomena of planetary disturbances, therefore they must also characterize the Jovial perturbation. Hence we know that, *a priori*, that intensely cold weather is liable to occur during the prevalence of a Jovial perturbation. We wish it distinctly noted that what we say does not imply that such weather is *inevitable*, but only that it *is liable* to occur. We have already explained the reason for this, by the statement that the temperature of the weather depends upon the locality where the prevailing high barometers during Winter originate. If they originate near or within the Tropics, then their downpouring and outpouring column of air is supplied by the equatorial permanently low barometer, and during the prevalence of such a barometer the weather will be unusually mild. But if they originate within the Arctic Circle, they will draw their supply of air from the upheaving column of the polar permanently low barometer, and the weather consequently will be intensely cold. Hence the winter temperature at all times depends upon the locality where the high barometer originated that for the time being is traversing a Temperate Zone. As during a Jovial perturbation the barometer is liable to act with extraordinary energy, abnormally cold weather is then likely to occur. Such are the deductions justified by the principles of the general theory of planetary perturbations; and hence we made an investigation to ascertain whether the facts of history verified the deductions. The following is the result of our investigation, and the correspondence or otherwise of all the facts we were able to find. It must be borne in mind that in history only the year is recorded; and as they are winter phenomena, we are unable to say whether the events occurred at the beginning or the end of the year. Of course this prevented us from comparing them with any other equinoxes than those of Jupiter and Saturn.

In the year 401 the Euxine was frozen over 20 days. Jupiter's major equinox occurred in January, 401.

From October, 763, to February, 764, the weather was so intensely cold at Constantinople that the two seas were frozen 100 miles from shore. A Jovial equinox in 763.

In 1035 a frost occurred in England in Midsummer-day, "that destroyed the fruits of the Earth." A Jovial equinox in July, 1035.

In 1063 the Thames was frozen over for 14 weeks. Jovial equinox, 1064.

In the year 1076, dreadful frosts in England from November to April. A Jovial equinox occurred in December, 1076.

In the year 1294, the Cattegat was entirely frozen over. This frost does not fall within the limits of either a Jovial or Saturnian period.

In the year 1323, the Baltic was passable for travelers for six weeks. If this was at the close of the year, it comes within 17 months of a Jovial equinox.

In the year 1402, the Baltic was frozen from Pomerania to Denmark. Jupiter's equinox 1403.2.

In 1407, all the small birds perished with cold in England. Jovial equinox 1409.13. If, therefore, this frost occurred at the close of the year 1407, it was within 13 months of a Jovial equinox, and hence within the period of perturbation.

In the year 1426, horsemen rode on the ice upon the Baltic, from Lubec to Pomerania. A Jovial equinox in 1427.

In 1433, the frost in Germany was so intense, "that all the fowls of the air sought shelter in the towns." A Jovial equinox occurred in 1432.92. In England the Thames was frozen over below London Bridge to Gravesend, from November 24th, 1433, to February 10th, 1434. This event falls within the perturbation of 1432.92.

In the year 1460, the Baltic was frozen over, and horsemen rode across from Denmark to Sweden. A Jovial equinox occurred 1462.56. Unless this frost occurred at the close of the year 1460, it does not fall within the limits of the Jovial perturbation.

In the year 1468, the winter was so intensely cold in Flanders that the wine distributed was cut by hatchets. A Jovial equinox occurred in May, 1468.

In 1515, carriages passed on the Thames from Lambeth to Westminster.  A Jovial equinox occurred in December, 1515.

In 1544, wine froze solid in Flanders.  A Jovial equinox occurred in July, 1545.

In 1548, sledges were drawn by oxen on the Baltic, from Rostock to Denmark.  This event does not fall with a perturbation of either Jupiter or Saturn.

December 21st, 1564, diversions on the Thames commenced. In January, 1565, loaded wagons crossed the Scheldt. A Jovial equinox in June, 1563. These events took place from 18 to 19 months after, and therefore can hardly be included amongst the phenomena of the perturbation.

In the year 1594, the Rhine and the Scheldt were frozen over. A Jovial equinox in 1593.

In 1607, diversions and bonfires on the Thames. The nearest Jovial equinox occurred in November, 1604, therefore this event has no relation to a Jovial perturbation.

In 1622, the rivers of Europe and the Zuyder Zee were frozen over, and the Hellespont covered with ice. A Jovial equinox occurred in August, 1622.

In 1658, Charles X of Sweden crossed the Little-Belt over the ice from Holstein to Denmark with his whole army, horse and foot, with large trains of artillery and baggage. A Jovial equinox occurred in May, 1658.

In 1683-84, a terrible frost occurred in England. It began at the beginning of December, 1683, and lasted until the 4th of February, 1684. The Thames was covered with ice eleven inches thick; nearly all the birds perished; the forest trees, even the oaks, were split by the frost, and most of the hollies were killed. The nearest Jovial equinox to this event occurred in August, 1681, consequently the event did not occur within the period of a Jovial perturbation. A Saturnian perturbation however was prevailing at the time.

In the winter of 1691-92, the severity of the winter and the intense cold drove the wolves into Vienna, where they attacked cattle, and even men. A Jovial equinox occurred in May, 1693.

Three months frost and heavy snow occurred from December to March, 1709. This event does not fall within a Jovial perturbation.

A fair was held on the Thames, commencing on the 24th of November, 1715, and continued to February 9th, 1716. This event hardly falls within the Jovial perturbation, which culminated in July, 1717. A Saturnian equinox however took place early in January, 1716. The reader will remember that in February, 1716, those extraordinary auroras took place that were called "Lord Derwentwater's Lights," from the fact that one occurred on the day of his execution.

In 1740 a protracted frost occurred in England, that lasted nine weeks. Coaches plied upon the Thames, and all kinds of festivities and diversions took place upon the ice. This was called the "Hard Winter." A Jovial equinox occurred in April, 1741.

From December 25th, 1765, to January 16th, 1766, and from January 18th to 22d, a frost with the most terrible effects prevailed in England and Europe. A Jovial equinox occurred in January, 1765. This is the first instance where we have specific dates that enable us to test the accuracy of the deduction by comparing the dates of specific phenomena with those of the equinoxes of the inferior planets. Venus passed her equinox December 22d, 1765, being most probably accompanied by a low barometer with mild weather and a storm of more or less energy. This low barometer was followed by an Arctic high barometer, that with persistent pertinacity maintained itself until it was displaced by a low barometer evoked by the Mercurial equinox of the 16th of January. Two days after, another Arctic high barometer supervened, enduring only four days, when it was displaced by the low barometer attending the Vulcanian equinox of January 25th.

A generally severe frost prevailed over Europe in the Winter of 1788-89. The Thames was passable on the ice opposite the Custom House, from November to January. A Jovial equinox occurred about the 1st of October, 1788. As neither specific facts nor dates are given, we cannot apply the test of the perturbations of the interior planets. A Mercurial equinox, however, it may be stated, occurred on November 4th, and another on December 18th, 1788, and one of Venus on the 17th of January, 1789.

In England, an intense frost prevailed from the 24th of December, 1794, to the 14th of February, 1795, with but one day's thaw, January 23d. The Jovial equinox occurred on September 1st, 1794; Venusian equinox, November 25th; Mercury, December 23d; Vulcan, December 24th, 1794, and January 18th, and February 8th, 1795; Mercury, February 5th.

Intense frosts all through December, 1796; the 25th is said to have been the coldest day ever felt in London. Mercury's equinox occurred November 24th; Vulcan's, December 2d and 25th.

The cold in Russia in 1812, surpassed in intensity that of any winter for many years. It was very fatal to the French army in its retreat from Moscow. Napoleon commenced his retreat on the 9th of November, when the frost covered the ground, and the men perished in battalions; the horses falling by hundreds along the roads. With the loss in battle, and the loss of this terrible and calamitous frost, France, in the campaign of this year, lost 400,000 men. Jupiter passed his equinox on the 8th of June, 1812. The days of intense cold are not specifically named, only that the weather was intensely cold, from November, 1812, to February, 1813. The following were the planetary equinoxes about this time: Venus, February 12th, 1813; Mercury, October 22d, December 5th, 1812, and January 18th, 1813; Vulcan, November 20th, December 13th, 1812, and January 5th and 28th, 1813.

On January 13th, 1810, quicksilver froze hard at Moscow. The equinox of Venus occurred December 17th, 1809; that of Mercury, January 11th, 1810. It seems to have been an Arctic high barometer following the low barometer of the Mercurial equinox. In Norway, on January 2d, 1849, quicksilver froze. Jupiter's equinox had occurred just about a year before; Vulcan's, December 29th, 1848; Venus, January 10th, 1849. It seems to have been an Arctic high barometer following the Vulcanian equinox that probably had been accompanied with a low barometer and storm.

On the morning of the 3d, 4th and 5th of February, 1856, the thermometer in St. Louis County stood respectively 17°, 22° and 8° below zero; the weather then moderated somewhat, but it snowed nearly every day for the balance of the month. The equinox of Venus occurred February 7th. On the last days of January it had thawed and rained a little, winding up with a snowstorm on the 1st of February. The low temperature unquestionably was brought by an Arctic high barometer following the low barometer that had prevailed a few days before. It will be observed the cold high barometer preceded the Venusian equinox several days. This I find is almost invariably the case with all the planetary equinoxes. All that is exceptional in this case, is that the greatest intensity of cold came before, instead of after the equinox which is generally the case. The alternation seems to be first a high barometer then a low one, generally a day or so previous to the equinox, then almost invariably an Arctic high barometer follows with intensely cold weather during its presence.

From the 23d to the 30th of December, 1860, the cold was excessive in England. On the 25th of December, in some parts the thermometer fell as low as 20° below zero. Jupiter's equinox had occurred about thirteen months before; Mercury's occurred 21st of December, or two days before the excessive cold commenced; and that of Venus was about to occur, namely, on the 9th of January, 1861. This seems to have been an Arctic high barometer following the low barometer that attended the Mercurial equinox.

The Venusian equinox of January 9th, 1857, was immediately followed by an intensely cold spell of weather that lasted about two weeks, in which the thermometer marked at sunrise from 8° to 20° below zero.

Again I find the material collected too vast to be handled. Pursuing a line of thought that seems never to have occurred to any before, I had to feel my way cautiously, and at every step to assure myself that I stood upon firm and impregnable ground. This was a difficult task, since for this purpose I must have individual facts; and but few individual facts—they are, however, of the strongest marked characteristics—have escaped being swamped in the Dead Sea of averages. But the few, the very few that have escaped this fate are fragmentary; that is, they are not consecutive, or a concatenation of facts, showing what events preceded or followed them. For example: our theory postulates that at planetary disturbances, the following to be the order in which phenomena succeed each other: (1) a high barometer, and in Winter accompanied by severely cold weather, precedes the equinox. (2) A low barometer at the equinox, consequently moderate weather, accompanied by a rain or snow storm; and (3) A high barometer following the low, generally accompanied by extremely cold weather. To verify and demonstrate these assumptions we want all the phenomena, and in the order of their occurrence. Instead of having the whole order, we have fragmentary parts of it. If the phenomena of the first high barometer have been exceptionally severe, we have them alone, and consequently disconnected. If the phenomena of the low barometer have been extraordinary, they are noted, but no statement of the events that either preceded nor followed them. If the phenomena of the second and last high barometer have been remarkable, we find them recorded, but nothing is said of the attending conditions nor circumstances. Moreover, the facts stated are mere skeletons of the events that have taken place. For instance, if intensely cold weather has prevailed, the material fact is not stated that it was the concomitant of a high barometer, for a few full statements would soon suggest the inference that the two phenomena stand to each other in the relation of *cause* and *effect*. Hence the omission is vital. This is precisely the defective condition facts are found to be in, by every investigator who makes original researches, and who attempts to trace through consecutive phenomena the operations of a physical law.

We have any number of extraordinary hyemal phenomena, such as excessive snow falls, and intensely fierce frosts; but all we can do towards a verification of the theory is to produce the astronomical testimony that one kind have occurred at a planetary equinox and the other kind has immediately preceded or followed such an equinox. This is very strong circumstantial evidence, but not positive proof. We will present a few of these phenomena in verification of the theory, where we have the continuous observations, or where we have been able to exhume the missing facts; and then we will close with the hope that the mode of recording and preserving observations of physical phenomena may be changed in the Future, so that when any physical law is discovered, it can be traced out, demonstrated and verified by the records of facts; for no physical law can now be specifically established by observations in the Past.

The first phenomena we will adduce are those whereof we

have the consecutive facts as observed by Sergeant Charles P. Fish, at the station, Island of Saint Paul, Behring Sea. A Vulcanian equinox occurred on the 24th of January, 1873. Mr. Fish's observations are, that from January 17th to 19th, a high barometer, 30.19 at times prevailed, and for the first time during the season the thermometer was below zero; on the 18th it was 10° below. Under a falling barometer, descending as low as 29.32 on the 25th, the thermometer rose to 20° above zero. Under a rising barometer attaining to 30.40, the thermometer again fell to 8° below zero, on the 27th. The facts in order are (1) High or rising barometer from 16th to 19th, with temperature below zero. (2) Falling or low barometer from 20th to 25th, (the equinox on the 24th) temperature above zero; and (3) Rising or high barometer from 26th to 28th, with temperature again below zero.

The next phenomena are in all respects similar, only modified by the fact that Vulcan's equinox of February 16th, 1873, was complicated with a Mercurial equinox of February 18th. The barometer at the same station, on the 8th day of February, stood at 28.82. It then commenced rising, and on the 10th the thermometer stood 12° below zero; it remained below zero all the time from the afternoon of the 9th until noon of the 13th. With a falling and low barometer (28.44) on the day of the equinox, the thermometer had risen to 34°, with rain and snow from the 13th. The Mercurial equinox on the 18th was accompanied with a low barometer, (28.32), mild temperature, and rain and snow every day until March 1st, when a high barometer (30.31) sent the thermometer down to 5° below zero again. At the Vulcanian equinox on the 11th of March, the observations at the same station show a repetition of the same phenomena, namely, March 9th and 10th, thermometer below zero; March 11th and 12th, time of equinox, above zero; March 13th to 17th, thermometer below zero. These three are the only instances we were able to find in which there are consecutive observations. It will be seen they indicate the order of the appearance of the phenomena to be as we have stated it.

The following instances, though the facts are incomplete, we present as a further illustration of the principle:

A condensed statement of the observations at the station at

Breckenridge, Minnesota, for December, 1873, is as follows: December 2d to 6th, from $2°$ to $25°$ below zero. 7th and 8th, from $8°$ to $11°$ above, (Venusian equinox on the 9th.) 9th to 14th, $8°$ to $10°$ below. Vulcanian equinox on the 12th. 15th, $1°$ above. 16th, 0. 17th to 25th, from $9°$ to $29°$ below. Mercurial equinox on the 23d. 26th to 31st, $8°$ above to $9°$ below. January 1st and 2d, 1874, from $15°$ to $20°$ above. From the 3d to the 6th, from $8°$ to $29°$ below. Vulcanian equinox on the 4th. The order of the phenomena were, high barometer and cold appeared seven days, and low barometer and rise of temperature two days before the Venusian equinox; followed on the day of the equinox by a high barometer, which extended to the 14th, retarding the Vulcanian low barometer three days; which only made itself felt for two days; then followed the severe cold of a high barometer, that retarded the Mercurial low barometer three days. After this low barometer followed a high barometer, that for two days sent the mercury below zero; on January 1st and 2d the Vulcanian low barometer appeared, being accelerated two days by the Mercurial and Venusian influence yet prevailing. Then came a high barometer on the 3d to the 5th, with thermometer from $8°$ to $23°$ below zero; followed by a second Vulcanian low barometer, with temperature from $3°$ to $12°$ above from the 6th to the 9th. To this succeeded on the 10th a high barometer that continued to the 16th, with a temperature from 0 to $23°$ below. Another Vulcanian equinox occurred January 27th. The high barometer that preceded it on the 24th, sent the mercury $33°$ below zero at Breckenridge. The low barometer at the equinox brought the mercury up only to $0°$, and on the 29th it fell again to $16°$ below, under the high barometer,—the highest of the month—being at Pembina, 30.98 inches.

A Vulcanian equinox occurred on the 19th of February, 1874. At Breckenridge, from the 11th to the 17th, the temperature ranged from $3°$ to $25°$ below zero. With the Vulcanian low barometer it rose to $23°$ above on the 18th. Under the high barometer that followed, it sank to $29°$ below on the 25th. On the 24th the thermometer stood at $-24°$ at Cheyenne; $-17°$ at Colorado Springs; $-9°$ at Denver; and at zero at Santa Fe.

A Vulcanian equinox occurred on the 7th of January, 1875. Of the low barometer succeeded by the high barometer on the 8th, we have already spoken; and also of the accompanying earth currents. The temperature on the 8th and 9th, at the following stations, was: Breckenridge, 8th, —31°, 9th, —33°; Cheyenne, —23°, —38°; St. Louis, 21° on 8th, —15° on the 9th; Denver, —29° on the 9th—the lowest temperature ever observed there. Mercury's equinox occurred January 23d. The high barometer that preceded it on the 13th, 14th and 15th, sent the mercury down to —30° at Pembina; —34° at Breckenridge; —17° at Omaha; —8° at Chicago; and —2° at St. Louis. Then followed the low barometer of the Mercurial equinox, whose accompanying storms have already been presented. This low barometer was followed by high barometers IX and X of the Signal Service, producing, on the 27th at Mount Washington, the extraordinary low temperature of 45° below zero. The Vulcanian equinox on the 30th was accompanied by a low barometer on the date of its occurrence, which was succeeded on the 2d and 3d of February by another extremely low barometer, 29.10 inches, followed on the 6th by the extraordinary high barometer of 30.98 inches. The Weather Review does not give the temperature of this high barometer, though it gives a list of the low temperatures reported—some as low as —48°; yet as it does not give the dates of their occurrence, we cannot verify so much of the theory as postulates that of two high barometers coming from the Arctic region, the one showing the greater pressure, being the more energetic, will hence effect the greater change in temperature. This, like many other points raised, has to be determined by future observations.

It is almost impossible to state a general principle so fully that a captious critic cannot find facts that he supposes contradicts it. The earnest, impartial and candid investigator too, is sometimes startled and perplexed by meeting with facts that he is unable to reconcile with general principles that he has verified to be true. In regard to that part of the theory relating to hyemal phenomena, we have endeavored and will endeavor to enable any one who candidly wishes to attain the truth, to overcome all perplexities that may occur. In the journals in which I first called public attention to planetary cycles, I stated in general terms the principles of the theory, and the manner the Atmosphere is affected, both in Summer and in Winter. Amongst the facts stated were, that generally the effect in Summer was excessive precipitation; and in Winter a deficiency. Captious critics immediately called upon me to reconcile this—as they styled it—contradiction. In the simplicity of their hearts they had never risen to the conception of the idea that the economy of Nature is conducted upon the principle of alternation, though in heat and cold, in Summer and Winter, in sunshine and rain, in day and night, the lesson is inscribed upon the Earth and Sky. They hence failed to conceive that the moving force of the Universe, Electricity, must have alternate states also, that is, it must be alternately static as well as dynamic.

Their error was in conceiving it to be continuously dynamic. We think that enough has been said to enable every one capable of perceiving the difference between a static and a dynamic condition, not only to see how these apparently contradictory results harmonize with the theory, but the principles by which these results are effected and the absolute necessity for them.

An apparent contradiction will also frequently, and we may say always, be observed when the energy and temperature of high barometers are compared, without regard to their sources. A comparatively feeble Arctic high barometer will pour out air of far lower temperature than the highest barometer coming from towards the Equator. The principle will only hold when the temperature of one Arctic high barometer is compared with that of another; and the principle is mainly stated for the purpose of drawing attention to the fact that, it is the Arctic high barometers that pour the intensely cold air over and upon continents. The same principle obtains in the southern high barometer that pours out in Summer the hot and desiccating simoom on the southwestern Plains, as we have already stated. On the Plains in southeastern Colorado, by personal observation, I ascertained that this simoom had a temperature of 105° Fahrenheit.

In comparing the temperature of one continent with that of another, at the same season, frequently opposite conditions are found to prevail. For instance, in America the Winter may be mild, and in Europe unusually severe. Müller, in his *Kosmischen Physik*, shows that the temperature at Berlin, in December, 1829, was 16°.6 Fahrenheit below the December average for that locality. In Petersburg, during the same time, though the temperature was below the average, yet it was much less so than at Berlin and Paris. In Irkutsk, Siberia, it was considerably above the average of December for that locality. The same was the case in Iceland and in America. Upon what principle, consistent with the theory, can these seemingly contradictory facts be explained? It is evident that during that time an Arctic high barometer was pouring down upon Western Europe a constant stream of cold air, probably supplied by the polar low barometer, or probably the one in Siberia; for since the temperature of Siberia was above the average, it either had a low barometer, or else a high barometer—from a more southern latitude—had swung so far North as to cover Siberia, which is not probable; for then the Asiatic and European high barometers, mutually attracting each other, as all similar barometers do, would coalesce, and one would be obliterated. In Iceland, a dynamic condition prevailed, because it is normal there in all seasons of the year. During this season it probably drew its supply of air from southern latitudes. In America the condition was like in Siberia, either dynamic or static. If the latter, then the prevailing high barometers came from the South. But it is not probable that a static condition, that is, a downpour of air prevailed for so long a time simultaneously over three continents in the same zone. That there was a high barometer or down-pour of cold air on the continent of Europe during the period, is incontestable. But as we have said before, whenever there is a down-pour, there must also, somewhere in the vicinage, be an up-pour to carry off this down-pour. The probability therefore is that, either Asia or America, or both, were pouring up their superincumbent air; while Europe was drawing it down. The only thing extraordinary about this phenomena is the continuance and protraction of these states, without material change for so long a term, in the same localities. These however are questions of minor importance, and are only mentioned because they may be stumbling blocks to those not yet confirmed in the faith "that maketh wise unto intellectual salvation." Other questions of weightier import loom up that deserve our earnest attention, and that will for several generations tax to the utmost our mental resources and all of our fund of Knowledge for their solution.

We have spoken of the perturbations of the Earth and the Atmosphere in consequence of having their electric condition intensified by a periodical augmentation of the electric tension of one or more other planets; and we have demonstrated that the consequences resulting from this augmentation are physical paroxysms in both the Earth and Atmosphere. What purpose does a paroxysm answer? Or is it without a purpose? Does it relieve the Earth and the Atmosphere from an abnormal electric tension; or does it leave them in the same condition they were in before? If it does, then it is superfluous; and Nature is at fault and defective; for here she has incorporated in the structure and government of the Universe an element of discord that, instead of being a means of conservation, is not only one of destruction, but of purposeless destruction. Who can believe that Nature is guilty of such an absurdity, not to say, folly?

These periodical flows of Electricity and consequent paroxysms were ordained for some wise and beneficent purpose. They not only quicken the Earth from its centre to its circumference, but they eventuate in spasms that throw off the now effete force.

But the Earth is not the only planet in the Solar System; it is only one of a group. Now, as a member of a group, whatever affects it, affects all the rest; and whatever affects any other member, affects it; for in a group of electrified bodies the electric tension on any one can neither be augmented nor diminished without affecting a corresponding modification of the electric condition of all. If the Earth relieves itself by a paroxysm, then instantly every other planet must do the same, or else the equilibrium of the Solar System would become unbalanced, which would be attended with most direful consequences, ending in a general catastrophe. Hence if it be the Earth that is first seized with a paroxysm, all the other planets must follow suit; or if the paroxysmal crisis first supervenes on another planet, the Earth instantly catches the infection. Each planet is, as it were, distended with an effete element,—that is with Electricity exhausted of all its ministering and nourishing properties,—and with the usual spasmodic attack imminent, that brings relief. Now it may be the Earth or some other planet that is first attacked, but the infection almost instantly spreads to the whole group. It may take several days, or even a week or more, for it to run its course; hence several paroxysms generally follow each other at short intervals during the crisis imposed by a planetary equinox. Suppose the Earth throws off first, and no other planet responds immediately, then the paroxysm on the Earth will not run its full course, but will be checked up and temporarily stopped by the controling influence the unmoved charges on the other planets exert upon it. But as one planet after another throws off, the last planets or planet that does so, will exhaust its charge to the extreme limits permitted by the charges on the other planets. By this time the Earth is far more out of electric equilibrium than before, and another and a more violent paroxysm ensues than the first. Other planets follow, and so on indefinitely until equilibrium in the system is once more established. This is the true cause why, when there is extraordinary high electric tension,—as there always is during the prevalence of a Jovial perturbation,—the paroxysms are so protracted, for they must continue until the cause of them is exhausted.

In the Earth and in the Atmosphere, all phenomena are inseparably connected with high and low barometers. Even earthquakes are not an exception. Down-pours and up-pours, or an interchange of air between the surface and the bottom of the ærial ocean, are the means by which a planet receives and gives off energy. Analogy leads us to suspect the same to be the case with the Sun. Sunspots and vast protuberances are phenomena synchronous with planetary equinoxes. Secchi observed and drew a sunspot on the 5th of May, 1857, that exactly represents a gigantic whirlpool, or funnel, into the interior of which the photosphere appears to be rushing with an eddying motion. Photographic pictures of sunspots, taken by De la Rue, when placed side by side, and looked at through a stereoscope, exhibit exactly the form of a funnel. The inevitable inference therefore is, that sunspots are of the nature and character of the down-pouring vortexes of high barometers in our Atmosphere. While the drawings made by Young, Zöllner, and others, of protuberances, so exactly resemble the up-pours of the cyclonal vortex under a low barometer, as to leave no doubt of their identical character. But we have fully discussed this subject in Part

I. We only call attention here to the fact that the Sun suffers similar paroxysms as the planets do at these equinoctial disturbances.

Truth is always prolific. The higher our standpoint on scientific truth, the more enlarged and comprehensive is our view of the Universe. Many objects—though being upon higher ground, yet hidden before—of the highest interest and importance, now loom up and beckon us to come up higher. Complying with their invitation, we soon find ourselves on an immeasurably higher elevation, with views proportionally enlarged, yet disclosing still higher ground. Such upward progress can be pursued indefinitely, for as the mysteries and resources of Nature are our subject matter, we can never exhaust it. Hence it is impossible for us to set bounds to Human Progress in Science in the Great Future that is before us. But this progress is conditioned, that we lay every step in it upon the immovable bed-rock of Truth.

In these paroxysms of planets we have seen they receive, through the down-pouring column of air, energy from Space, which they return through the upheaving column, as soon as it has performed its functions in the economy of the planet. All electricians know that Electricity can only pass from one point to another by one of two ways, namely, Conduction where Matter is continuous, and Convection when it is discontinuous. The problem, therefore, confronting us is: If it be Electricity that comes from Space and returns thither, what gives it Convection? We direct attention to this problem, not with a view of discussing it, that we have already done elsewhere,* but to point out that the line of thought we are pursuing, must ultimately embrace in its scope the condition of Space, and settle the mooted point whether it is a vacuum or a plenum. It will do more. It will show the interaction and relation between Space and the orbs which are sparsely disseminated throughout its boundless realms.

## CONCLUSION.

When starting to explore the immense field traversed by us,

---

*Note.—In Part I, and more especially in a work in preparation, giving the cause of Meteoric Showers; and an astronomical demonstration of that cause and a verification of it by historical facts.

we took for our sole guides the known laws and causes operating in and upon our own Globe, as far as we could apprehend them by contemplating physical facts inscribed upon the Earth and upon the Sky. At no time, either by day or night, and at no point on our route, however indistinct and obscure, have they failed us; although we were surveying unknown fields, traversing unexplored regions, and were surrounded by strange scenes and unfamiliar objects while pushing a reconnoissance to the verge of the Universe, where abysms gaped wide, whose depths no mortal ken can pierce, nor human reason fathom. The greatest and most imposing castle erected by Error in the Past, we have seen crumble and disappear like a dissolving view in a phantasmagoria. We have discovered that neither the Solar System nor the Universe is a Mechanism, passively obeying commensurable and mechanical laws; but that both are symmetrically constituted, and most delicately adjusted systems of worlds, affected by polar forces operating with omnipotent energy, yet each member inherently contributing to the preservation and highest welfare of all, while securing the stability and permanency of the whole. We have seen that taking either the Solar System, or the Universe, though its members are distributed over immense distances in Space, yet it is an indissoluble Unity, bound together by such intimate ties and sympathies, that whatever affects one member, sends a thrill that vibrates through the whole system. We have pointed out this mysterious bond of union and sympathy; and in the eclaircissement of physical phenomena, from the most obscure, indistinct and inobtrusive, to the most conspicuous, wonderful and imposing, we have unfolded its laws so plainly and clearly that "he who runs may read," and understand, and comprehend the most amazing, beautiful and complicated problems proposed, in the structure of the Universe, for Man's admiration, solution and inspiration.

While we claim to have done this, yet we have by no means exhausted the mysteries surrounding us on every side, that stare us in the face in every direction we glance, and are discovered at all points to which we fix attention. Nay, we have hardly trenched upon their borders. At most the investigation has revealed only what was partially concealed before; and directed attention and inquiry to points hitherto passed by and neglected.

New orders of facts, new causes and relations, extending into the illimitable depths of Space, have been disclosed, and we have endeavored to give their most obvious interpretation.

As we advanced, wider fields for observation and exploration opened before us, tempting us to enter and possess them, but we had to exercise self-denial and leave them to future generations to appropriate, survey, occupy and cultivate. We have contented ourselves with the mundane sphere; and have imposed upon ourselves the humble task of ascertaining and making a full inventory of all the facts occurring therein, with their variations and modifications, according to circumstances and conditions. Such a task is the prose of Science, but at the end of it, comes inspiration and poetry. We have indulged Fancy but sparingly; at most only to try her unfledged wings in short flights, not much above the level of the common place, because in our enthusiasm we could not help it; but we have rigorously not permitted her to pass the boundary of clear intuition, circumscribed by the logical deductions of Reason. Whenever our inventory of facts relating to a given point were full enough to justify it, we have made deductions of all the laws and causes warranted by the facts, of the physical phenomena in question. Nor did we leave the matter here, but we proceeded and demonstrated the truth of the deductions thus made by incontestable facts. We have literally and rigidly obeyed the injunction of Locke, "In Science write demonstratively," and have endeavored to lay the foundation of the Temple of Science upon the immovable bed-rock of facts. The walls of the edifice will rise dazzling with beauty and resplendent with Light, as fast as compact and cemented facts, the necessary material for its erection can be collected. In the day of its triumphant completion, we hope our humble labors will be worthy of so much recognition as to be estimated as a handful of sand gathered on the shores of the Ocean of Time, and contributed to the great end.

# APPENDIX.

# APPENDIX I.

## PLANETARY EQUINOXES FROM THE YEAR 1866 TO 1884.

Vulcan (o).  Mercury (O).  Venus (*O*).  Terrestrial (not given).

| Months. | 1866 | 1867 | 1868 | 1869 | 1870 | 1871 | 1872 | 1873 | 1874 | 1875 | 1876 | 1877 | 1878 | 1879 | 1880 | 1881 | 1882 | 1883 | 1884 |
|---|---|---|---|---|---|---|---|---|---|---|---|---|---|---|---|---|---|---|---|
| Jan. | 3 / 9 / 26 | 6 / 29 | 9 / 27 | 8 / 12 / 13 | .... / 15 | .... / 18 / 31 | 18 / 21 | 1 / *5* | 4 / *8* / 24 | *7* / 27 | 10 / 10 / 30 | 8 / 13 | .... / 16 / 27 | 14 / 19 | 1 / 22 | 2 / *8* / *25* / *31* | 5 / 18 / 28 | 5 / 8 / 31 | .... / 11 |
| Feb. | 18 / 22 | 9 / 21 | 1 / *5* | 4 / 26 / 24 | 7 / 13 / 27 | 10 | *5* | 16 / 13 | *5* / 18 / 19 | .... / 22 | 2 / *5* / 22 / 25 | *5* / *9* / 28 | 8 / .... | 11 / 27 / .... / 14 | *5* / 14 / 17 | .... / 20 | .... / *23* | 18 / .... | 3 / *4* / *5* / 26 |
| Mch. | 13 / .... | *5* / 16 / *25* | 12 / 19 | .... / 22 | 2 / *25* / *29* | *5* / *5* / 16 / 28 | 3 / 8 / 31 | .... / 11 | *5* / 14 / *21* | *5* / 8 / 17 | .... / 20 | .... / *23* / *25* | 3 / *12* / 26 | *5* / 6 / 29 | *9* / 30 | .... / 12 / *16* | 3 / 15 | *5* / 18 | 20 / 21 |
| Apr. | 2 / *5* / *7* / 28 | 8 | 11 / *25* | 11 / 14 / .... / *30* | 2 / 17 | .... / 20 / *29* | 16 / 23 | 3 / *3* / *26* / *30* | 2 / 6 / 29 | .... / *9* / 21 | 7 / 12 | .... / 15 / *30* | 2 / 18 / *25* | 12 / 21 | 1 / 24 | 4 / 27 / *29* / *30* | 2 / 7 / 16 / 30 | 3 / 10 | .... / 13 |
| May. | 21 / 21 | 1 / 8 / 24 | 4 / 27 / *28* | 7 / *25* / 30 | 10 / 12 | 13 | 16 / *28* / *30* | 17 / 19 | 4 / 22 | 2 / 25 | *5* / 21 / *28* / *28* | 8 / 8 / 31 | 11 | .... / 14 / 26 / *28* | 12 / 17 | .... / 20 | .... / *23* / 30 | *3* / 17 / 26 | 3 / 6 / *27* / 29 |
| June | 13 | 16 / 21 / *25* | 8 / 19 | .... / 22 | 2 / *25* / *25* | *5* / *12* / *25* / 28 | 8 | 11 / .... / *30* | 14 / 17 | 4 / *17* / *25* | .... / 20 | 21 / 23 | 3 / 8 / 26 | 6 / *25* / 29 | 9 / *25* | 12 / 12 | .... / 15 | .... / 18 / *25* / *30* | 16 / 21 |
| July | 4 / 6 / *23* / 29 | 9 | 12 / 22 | 8 / 15 | .... / 18 / *23* | 21 / 26 | 13 | 4 / 24 | *7* / 27 / *30* / *31* | 10 / 18 | 4 / 13 | .... / 16 | 19 / *22* / *23* | 9 / 22 | 2 / 25 | *5* / 26 / 28 | 8 / 13 / *23* / 31 | 11 | 14 / *30* |
| Aug. | 17 / 21 | 1 / 4 | 4 / 24 | 7 / *20* / 21 / 30 | 8 / 10 | 13 | 16 / *26* | 13 / 19 / *26* | .... / 22 | 2 / *25* / 31 | *5* / 17 / 28 | 8 / *20* / 31 | 11 | .... / *22* | 14 / 17 | 8 / 20 | 20 / *26* | .... / *23* / 26 | 3 / 13 / 29 | 6 |
| Sept. | 13 / *30* | 16 / 17 | 4 / *18* / 19 | .... / 22 | 2 / 21 / *25* | *5* / *8* / 28 | 8 | 11 / *18* / *26* | 14 / 14 | 17 / .... | *18* / 20 / *30* | 17 / 23 | 3 / 4 / 26 | .... / 6 / 29 | .... / *9* / *18* / 21 | 8 / 12 | .... / 15 / *26* | .... / 18 / 21 | 12 / *17* |
| Oct'r | 6 / 29 | 9 / *15* / *31* | 12 / 18 | 4 / 15 | .... / *15* / 18 / 21 / *22* | 1 / 9 / 24 | 4 / 27 / 30 | *7* / *15* | 10 / 14 | 13 | 16 / .... / *31* | 18 / .... / 19 / 22 | *5* / *15* / 25 | 2 / *22* / 28 / 31 | 8 / 9 / *15* | 11 | 14 |
| Nov. | 13 / *14* / 21 | 1 / 24 | 4 / 27 | 7 / 17 / 30 | 4 / 10 / *14* | .... / 13 | 16 / .... / *22* | *9* / 19 | *14* / 22 | 2 / *25* / *27* | *5* / *13* / 28 | 8 | 11 / *14* / 18 | 14 / 17 | 4 / 20 | .... / *14* / 22 / 23 | 3 / *9* / 26 | 6 / 26 / 29 |
| Dec. | 14 / 27 | 14 / 17 / .... | 1 / 20 / .... | *9* / 23 / *31* | 3 / 18 / 26 | *5* / 6 / 29 | 9 | *9* / 12 / *23* | 10 / 15 | .... / 18 | 1 / *9* / 21 / 24 | 1 / 4 / 27 / 27 | 7 / 30 | 10 / .... / *18* | *5* / *9* / 13 | .... / 16 | 19 / *23* | .... / 9 / 22 |

APPENDIX I—CONTINUED.

The TERRESTRIAL EQUINOXES occur on the 22d of March and September in each year.

MARTIAL EQUINOXES have occurred and will occur in the years embraced in the following table:

| | |
|---|---|
| November 19th, 1866. | April 12th, 1876. |
| October 27th, 1867. | March 21st, 1877. |
| October 5th, 1868. | February 28th, 1878. |
| September 14th, 1869. | February 6th, 1879. |
| August 23d, 1870. | January 15th, 1880. |
| August 6th, 1871. | December 24th, 1880. |
| July 9th, 1872. | December 3d, 1881. |
| June 17th, 1873. | November 11th, 1882. |
| May 27th, 1874. | October 20th, 1883. |
| May 4th, 1875. | September 29th, 1884. |

JOVIAL EQUINOXES, September 25th, 1871; August 1, 1877; and July 6th, 1883.

SATURNIAN EQUINOX November 12th, 1877.

DIRECTIONS FOR FINDING THE DATES OF EQUINOXES.

To find whether a given physical phenomenon occurred at or about a Jovial equinox, reduce the days and number of months to the decimal of a year; subtract the date thus reduced from 1871.74, or any of the Jovial periods given, and divide the remainder, 11.86, the length of the Jovial year. If there be a remainder, then reduce the decimal part to months and days, and it is the exact time the equinox occurred, after the event. If it be short a small fraction, then the equinox occurred that much before the event.

TO FIND THE DATE OF A VENUSIAN EQUINOX.—Suppose you wish to find whether a Venusian equinox had anything to do with the hurricane that destroyed Surat in India, April 22d, 1782, subtract 1782 from 1874, 1878 or from any other year given in the table that will give a remainder divisible by 4, then multiply 1.2 day by the number of tens in the remainder, and add the product to the date of the equinox in the year in the table taken, and it will give approximately the date of the equinoxes for the year sought. Thus, if the year 1874 be taken, then the remainder is 92 years, which is divisible by 4, and 9.2 times 1.2 day is

11.04 days, which added to April 2d, the day of the equinox in 1874, gives April 13th, 1782, as the date of the Venusian equinox. Hence it occurred 9 days before the hurricane. Exact calculation would bring it still closer.

TO FIND THE DAY OF A MERCURIAL EQUINOX FOR ANY PAST EVENT.—Suppose it is desired to find whether a Mercurial equinox had occurred near the time the ship *Earl of Moir* was wrecked, August 8th, 1821. Now, since on every 23 years, or on any multiple of 23, the Mercurial equinoxes occur on the same days: therefore add 23 to 1821 until it comes within the limits of the years in the table, 46 added to 1821 gives 1867, and Mercury's equinoxes occurred in 1821 on the same days they did in 1867. The table therefore shows Mercury's equinox occurred on the 4th of August, 1821, or four days before the disaster. The following table will assist in making calculations:

For 1 year add 13 days.  For 33 years add 2 days.
For 6 years subtract 8 days.  For 46 years subtract 7 hours.
For 7 years add 6 days.  For 105 years subtract 3 days.
For 10 years add 2 days.  For 125 years add 1 day.
For 13 years subtract 2 days.  For 217 years add $\frac{2}{3}$ day.
For 20 years add 4 days.  For 309 years subtract 5 hours.
For 23 years subtract 3 hours.  For 572 years subtract 2 hours.

The equinoxes of Vulcan are only approximations. As now provisionally fixed in every 23 years, they occur on the same days. By inspection of the table it is seen they advance three days every year going forward, and recede three days going backward. Hence, in counting backward, we multiply the years by 3, and subtract the remainder from the date taken, thus in ten years the equinoxes will all have receded 10 times 3 days, or 30; dividing which by 23 leaves a remainder of 7 days, which subtract from the date of the equinoxes in 1876, gives the equinoxes in 1866, and so on with any of the years taken in the table. This it will be seen is for calculating his equinoxes in the Past. To find those of the Future, we must reverse the order, that is, *add* where we *subtract*, and *subtract* where we *add*.

The Table to find Mercury's equinoxes also has reference to the Past; to find those in the Future, reverse the process described in the table, that is, *add* where the table says *subtract*, and *vice versa*.

# APPENDIX II.

### FORMS OF CLOUDS.

Meteorology takes precedence of all Physical Sciences in its constant and direct application to immediate use. It affects all conditions in life, and all employments and industrial pursuits. Hence all are interested in it and concerned about it, and must feel more or less desire to know what will be the character of the impending meteorological changes; and at what time they are likely to appear. We may know the astronomical epochs when these changes will recur, the causes that bring them about, the modes by which they are effected, and the laws that govern them; yet these do not fix the time with sufficient precision to make it at all times available for practical purposes. The reason why this is so is because there are so many adventitious causes constantly occurring, not yet subordinated to known and fixed laws, that accelerate or retard the development of phenomena. Additional knowledge is therefore necessary, which will enable us to determine the presence or absence of these adventitious causes, and to detect the incipient stages, and watch the progress of the development of the coming phenomenon.

One item in this additional knowledge is to be able to interpret the movements of the mercurial column in the barometer. This point is fully discussed in Part I. But above all it is necessary to be able to read and understand what is written on the clouds and sky, in the color of the latter and in the form and shapes of the former.

The formation of clouds can be studied most advantageously in elevated mountain regions, where they are daily produced under the eye of the observer. During several summers, while sojourning in the Rocky Mountains of Colorado, I had ample opportunities to watch and observe the whole process, from its incipiency, in the attenuated mist that in the early part of the day appears over every towering pinnacle of the Snowy Range, until it culminated in the thunder storm, and then its gradual wasting away until it finally sunk back and disappeared in the same pinnacle from which it sprung. As in Part I the whole process is described, step by step, and the laws and causes in which the phenomenon originates and is controlled, further remarks on this point would be out of place here.

Cloud formation has many modifications, but are generally classed into the following five primary divisions: (1) *Stratus;* (2) *Cirrus;* (3) *Cumulus;* (4) *Cirro-cumulus;* and, (5) *Nimbus.* All these species are represented in the engraving fronting the title page. There are other intermediate forms, representing the transition state between the five primary forms, which in a short time, by daily observation, are readily recognized and distinguished. A few remarks may be necessary to enable the reader to recognize and perceive the differences in these forms and to interpret their meaning.

The classification is that of Luke Howard, presented in 1802 Prof. Andre Poey has presented another classification, but which is too complicated for the general reader.

(1) The stratus is the *bed* or *covering* cloud, which, at all seasons,—but more especially in Autumn,—appears near sunset and rests all night, till after sunrise next morning, along the edge of the horizon. It is most frequent on the evening of a day when cumulus clouds have prevailed but vanished towards sunset. It seems to be formed of a kind of a mist or fog thrown up by the Earth, or that has settled down from the dissolved cumulus. and is positively electrified. Hence on the night it prevails little if any dew falls. It must be distinguished from what has been called the cirro-stratus. The former is a whitish grey, the latter is of a lead color, and when it prevails heavy dews fall. The prevalence of stratus clouds indicate approaching rain.

(2) The cirrus or curled cloud is so-called because it is frequently seen with recurved tufts or wisps. It is the highest of all clouds, and is either the outflow over the upheaving column of a low barometer, or the remnant or skeleton of a storm centre, after the latter has become exhausted. They generally precede rains, and when they enlarge and become dense they indicate copious rainfalls. When their edges are sharply defined

upon an *intensely deep blue sky*, they indicate heavy thunder storms.

(3) The cumulus cloud is readily recognized by its *heaped* or *piled* appearance, which is the meaning of the word *cumulus*. When detached cumulus clouds gather together they form what are called *stacken-clouds*, which is the transition state between the cumulus and the nimbus. The stacken-cloud rapidly passes into the nimbus. When stacken-clouds become capped with a hood-cloud, like mountains often are, or when the cirrus above them enlarges and increases in density, then heavy storms are near at hand.

(4) The cirro-cumulus cloud is intermediate between the cirrus and the cumulus. The sky when flecked with cirro-cumulus clouds, has received the common name "mackerel-backed." It always indicates an electric change as going on, and the near approach of a low barometer; consequently a rise in temperature. In Winter it indicates the breaking up of frosts and the coming on of rainy weather.

(5) The nimbus is the storm or rain cloud. It is the last stage of cloud formation, and passes away in the rain it pours down upon the Earth. It is intensely electric, in consequence of which in daytime has an orange hue by which it is easily recognized; and at night it is more or less luminous.

No rain ever falls, nor need be apprehended when a grey sky prevails. But the deeper the serenity and the more intensely blue the sky is, the more certain not only the occurrence of rain immediately, but of its copiousness. In such a state of the weather, not only is the margin of the cloud sharply defined upon the sky, but the involutions and convolutions of cloud upon cloud is equally well defined, and the clouds themselves seem smooth, as though composed of ice. Signs like these are unmistakable evidence not only of the imminence of storms, but of their energy.

www.ingramcontent.com/pod-product-compliance
Lightning Source LLC
Chambersburg PA
CBHW031828230426
43669CB00009B/1266